Eridan Ozma

"Prisoner of Eridan"
Book One of the
Urbis Phobia Series

By Tomas Londan
2015 by Amazon Inc.

Isbn: 978-151-904-0909

Contents:

Chapter One: Arrival
Page 4
Chapter Two: Memories Of Home
Page 45
Chapter Three: Gulf City
Page 74
Chapter Four: Zenobia of Eridan
Page 99
Chapter Five: The Mall
Page 134
Chapter Six: Fontana Entheos
Page 160
Chapter Seven: Trader Flavius
Page 184
Chapter Eight: Return to Terra
Page 220
Chapter Nine: Epilog
Page 258

"Arrival"

I spend a lot of time these days wondering how Mankind has survived so far, and for good reason. After months on another world in some distant star system I finally got over being just another stunned victim and took the idea of being held prisoner as an adventure, perhaps even positive in terms of personal growth. This story started with my car running out of gas at 3:00 am on a deserted highway near the suburbs of Cleveland, Ohio. Having no cash nor credit card, I tried to phone my friends to come over to rescue me. Recall standing in a hot, brightly lit phone booth on July 20th, back in 1978. Some garishly lit modern gas station stood dozens of feet away. T'was the only sign of humanity for miles, dwarfed by vast dark forest. Behind large windows could be seen a desk with cash register, snacks on racks, beer and pop in coolers, posters, stacks of batteries, plus ziggurats of oil cans, served by one lonely gas station attendant. Back in those golden days. Also recall one picnic table with pop machines, food machines and so forth sitting in a circle of light outside on their yard. We seemed to share an entire universe.

This scene was lit by glaring neon, with a vast expanse of forest beyond. Far overhead hung our clear bright Milky Way almost as if it was important to know which part of The Cosmos we happen to live in. As Bruce Springsteen once said in song, Darkness at the Edge of Town. Sweat soaked my tee shirt and bra. Salty liquid ran down my legs. Even my jeans got soaked. They were tight, unflared but still faded in style. Close by lurked my Chrysler as some dark, hulking beast of steel. My random call was making me feel guilt. I even thought of offering some favor to that station attendant for gas. (Sex?) That was another stupid idea. Just like taking that nice **38 snub nose** from my glove compartment, where most of them are, just to rob this poor man. Aside from some lamps up on poles running along the Interstate at long intervals, casting feeble yellow light - they even reminded me of dwarf stars - there was only one vast black void to be seen. It was as if my car, that phone booth, the gas station plus Route 90 were all floating out in Space.

Time passed. 3:20 by now. My phone rang forever at the other end. A certain piece hidden in my auto came to mind but I did not intend to **cowboy** this poor guy for a few lousy gallons of gas. Then my line was suddenly cut off. I became dizzy and felt my body sink downwards. My skin grew cold and tingled. Well, what substantial beings we are, my mind said. But then even my mind vanished. As my eyes closed, a star field rushed upward to expand into my face. Last recalled was some low buzzing noise as I became dead mass and felt as if I were sinking into the ground.

[Pause.]

I recall waking up in a strange room. Well, there are in fact many strange rooms in Cleveland I have awoken in... but not this one! I could tell at once that this was some totally alien place. This room itself was nothing but one large expanse paved with lime green shag carpet, white walls and low ceiling. One half of this room was platform under carpet with curved a rise of two feet. Even my walls had curved corners, which disoriented the human eye. All of this made for odd effects. Actually I was wrong. Had two small rooms to one side, which later we called my East wall. That was where the Sun rose but my main room was forty by thirty feet, and one wide side, along my South, was one long window of thick glass. This revealed a strange skyline half hidden by smog. Bottom half of this split level suite had two side doors of aluminum, which slid in and out of the wall automatically. Assume they were powered by sensors, which reminded of TV shows from 1965. That very era. Most of all **The Prisoner,** made in England.

On one side I found a modern kitchen made of aluminum, brass and stainless steel. This side room shared with my Main Room that same long window. (In fact, that dimension varied in this alien home.) As in, both bathroom & kitchen were two feet lower than the basic floor of my "main" room. Both also had floors of polished pink marble, lovely to look at. Both were the same size, and took up the entire East side. (Each one was fifteen feet square, making this entire suite thirty by 55 feet. Huge for bachelor. The entire cieling was normal white plaster like back home. End effect was to make us walk up steps into Main Room, then cross half of that to climb up a green slope into what I assumed to be "sleep" space. Since had no furniture, t'was easy to use carpet for that. This design prevented flooding from those two small rooms. Next I explored the WC. Had to use it anyway.

My WC had the same fixtures as any on Earth. It had one large vanity counter with mirror, lights and two sinks. Then came two toilet stalls and one large sunken bathtub. Again to prevent flooding. These fixtures were made mostly of lime green marble. Faucets, etc. of shiny brass. Searching inside my vanity & linen closet, found what we humans might need: Colgate paste for teeth, Ivory soap and shampoo. Fluffy pink towels. And so forth. This was like that hotel suite back in 2001. A Space Odyssey, only modern and real. Very American. Using my head, I guessed that this curvy design left extra space between floors for utilities. How advanced. This joint had to be soundproof. After my "piss" break I left my WC just to check out Main Room some more. Walls were made of painted white concrete, not flimsy plaster I could kick out. For you see, I had figured out ASAP that this suite was simply some kind of prison. Also it had to be on another star system. At that point I took one short snooze on the carpet. Total fatigue.

[Pause.]

Forgive my archeology tour. When next I was awake and off my floor, noticed I was soaked with perspiration to the max, leading me to assume that I had been teleported here. Did suffer tap water gladly because this dump was hot. Air was stale, but one push on a button gave me cool fresh air from vents above. Of course they had air conditioning. "Just hope this is not a nightmare as Mailer said," went my goofy mind. Have you guessed my racket yet? Yet one look outside revealed vast skyline, still impressive under its grey clouds. It was raining off and on. My name is Alice Roebuck from Cleveland, Ohio. Like, I am a Journalist by profession with B.A. in that subject. At age 25 in 1978 got my diploma. Mostly I can operate video recorders, cameras, cars, planes and various guns. Am not good in STEM subjects, but assertive and alert enough for my assumed job. Being from the USA in a decadent decade, we were much given to big cars, which is so cliche. And guns. I was about to make some outre comment about my new suite being "A condo in Miami" or something, but why bother? If you want to laugh or indulge in New Journalism or even just some hip humor, please go to elsewhere. (Actually, we enjoyed disco. Was it really that bad?) Now we are getting ahead of self. e are trying to report on UFO abduction here, as well as paint you some convincing portrait (or is it scape?) of an alien world. In 1978, much noise has been made about this Hi Tech, IC Chips and Silicon Valley action. My friends from back Home would help me with work on this case. Politics are now unrelevant. Yet suddenly, freaked out. Wasting time in this creepy place just speculating. Time and Space are what Life is all about. Right?

In shock, checked my watch. It was normal but did count out day, month and year. (Not digital.) Right then it was at High Noon, July 25, 1978. I had lost one week. How? Not that I could recall anything since last on Earth. Assume that we had to "jump" over here in stages. Orbits within orbits and even stars themselves racing past each other at variable velocity. It figured. Who are They? Faceless beings. To just relax, looked at decor. In dead center one fat marble pillar ran from top to bottom. It was also pink. Phone and power outlets, again like our own, studded it. Then my door. There was one. It was of thick stainless steel. No lock, since it was made to slide out from concrete. So. Sliding doors like on Star Trek. On one side at waist level was one small panel of steel with buttons. Needless to say, it refused to do anything for me. There was only one exit, which was unpossible to get out. Me and my Orwellisms.

The only reason I was into Future Shock is my job. Alvin Toffler was one of my fave reads as well as just university text. I tend to be au courant. Such as, this year, into that famous comedian Steve Martin. King Tut, eh? You can deduce from the above clue that, yes, we were superficial. My own "set" was so. What do I look like? Denim and work shirts along with winter boots or tennis shoes. Usually. What did often we smoke? Not always but on weekends. What Mr. Martin claims he did. Again, superficial. At this point, my mind was devoid of most woolgathering. As soon as junk from TV, music or books came in, such junk vanished. This alien world was really getting to me. Then hunger drove me to explore my kitchen some more. By the way, was still dressed the same as when we last left Earth. Soaked with brine. Smelly. Never mind bath. Let's do with some **nosh**. Ran for my fridge.

There I did not find what I half expected. (Blue gunk like **Dave Bowman** had to eat. Okay?) No, this was only some regular kitchen from Earth, judging by design. My counter was granite, which was for YUMPies only, but never mind that! Two steel sinks for dish washing. Cupboards full of china, glass and the usual needs. Cutlery in drawers. Waxed oak decor. Westinghouse for stove and fridge. Pink enamel. (Why pink?) Walls of ceramic tile. And food? Fridge held the same chilled and frozen food found in any market back home. Even down to brand names. In freezer was frozen meat and veggies. In one large closet were boxes of dry food stacked up high. As if I had to live from this supply for months. Items like sugar, cereal, cookies, and noodles. Even dried beans and canned food. There was not any alcohol but lots of coffee and tea. Even milk powder. Is this getting too technical? Suddenly I stood up with creepy vibes coming in. Something was picking my brain. Then we dropped my **Earthian** paranoia. Relaxed and made me some kind of meal. OK. Pasta with tomato sauce. Frozen tuna and Peek Freens cookies. Coffee made with warm stale water from their urban supply. At least we can boil it. That made me relax again. Then took off all my gear and went up into Main Room to sleep. What else to do? Before we pause, let me explain what this abode was like. Later on we found out that it was on the top floor of one gigantic hirise rental tower. The same one I now lived in.

Its name was Superbloc "334" on 334 Seaside Boulevard. That's in plain, good English, which some of Them speak. To no kind of surprise to me. My own enclave was Suite "6610" meaning it was on Floor Sixty. So each floor had hundreds of suites. Told you it was massive! From the outside you can see white marble and glass. Bauhaus. To be really alien, floors were not all for living. Behind my side walls stood parking nooks for private autos. (These came up in elevators.) Yet unknown to me that day, my Beast was there only meters away. Safe and sound over 700 feet over ground level. Keys left in ignition by my captors. Yet for now, let me rest.

Time to stop my report. Am writing this on "my" PC which was very much State of the Art back in 1978. It is now 2084. I am in an office run by NASA in downtown Houston, Texas. So why such an ancient "tech" method? Why not up to date? My captors decided, after an interval of months, to suddenly return me safely to Terra, as They call it. In the same way as we came to that Other Place which for long was unknown. No name. They allowed me to hook up an Apple II to modern 21st Century machines and such gizmos of their own. More REM came on. She dozed off.

BREAK> TOGGLE > BASIC, 1968; Of Ancient Man DOS.

MSDOS Language "BASIC" C:\> July 25, 1987. What happened is this: Alice Roebuck spent her first days learning to use this primitive PC. Its screen said "Cut any corny quotes that seem to be clever and get to your point."

"What do you mean?" I went in my head.

"Use this Low Tech machine for now. We can comlink with you. Sound familiar?" To me that sounded Buck Rogerish but why not? At least They do not appear in stupid gear like in our Golden Age serials. This is easy even for me. I can see this Apple or IBM junk in any home or office. Now show me your face. "As they said in the Gulf War Era, In

Fortress North Am, the opposite is true!" My alien said, "Unless we **interface** with your **machine** we may have to grunt at each other. No offence. It concerns semantics. And do not worry about corny anything. You may get that from us later on. Like, as some grunt for The Empire, we give you more shit later."

"What was that about?" **me says.** So The Other says, "No idea, Terran! You and I personally, as two organisms, have no fucking idea of what Karma and the Kosmos can serve us." Swear to **God,** I had to accept Her spiel on **Faith.** This bitch sounded like an She was an nerd, a phony, and not to be trusted.

[Pause.]

"After a long time spent exploring my suite I figured me was in a vacant but functional apartment complex sixty floors tall. Only exit appeared to be leading into..." so I typed. How did I get this machine? Was just sitting there on floor when me woke up hours later. Sun already low on its own horizon, which me say just to make my point about this other star system. That hot object up there was certainly not our own Sun. Aka Sol." Back to my story in plain English, they must have snuck this in on me in my sleep. Lucky me covered in cloth. There was also more stuff. Over near the kitchen were more wooden boxes of food. Even bottled water. There was one desk of wood plus chairs to sit on. Next to that one matrass covered with blankets. "Learn to think like we do. Then we can comply with more needs. As time goes on, you get more from us." Time: 5:30 PM on July 25. Same year. Like really? "Snarky are we?" as me aped ESP resonance of alien. And their speech.

C:\> DOS PROMPT. This place reminded me of hirises my buddies in Cleveland lived in. Our own Fortress Society. Crime wave 1967 to 1972, etc. Soylent Green as negative y overly depressing flick from 1973. That is when I looked at my door again and found this tiny panel on the right side at waist level. Useless. Script be alien, of course me could not work it. They never gave me the door code. For Suite #6610 in Superbloc 334.

A:\> More MSDOS notes. We wish to apologize as the local beings. Natives of Xiotan. Me Eridan. You American. Me have only some very few sounds and/or Units Of Meaning to communicate with Man like You. Me be "Eetee?" So that is what Hynek et Jacques Vallee said in books and videos. To which Alice thought, why do They mix words from French and Spanish with English? Faulty computers? Can They read my mind? To which the screen said "We certainly can and do read minds. Like your own. Just like in Ancient Hollywood." C:\> A logical proposition. Mucho bueno.

[Pause.]

Alice went on like this. Her style often imitates H.G. Wells. That was suggested by the ambient genre. She even knew factoids, such as, that writer was known as a Fabian Socialist. She often mentioned such concepts for effect. Yet what did the natives know about our history? Look at this room. What clues can it give? Do we suck up to leaders or poor of alien society? Well? Effect was indeed fortress like. Medieval yet modern. Clearly an alien room but in very subtle manner. It was clearly designed for Earth People, having that casual style called Bauhaus. Response?

"Blah blah and blah." quoth Other, "Try more like sterile and dull. Anonymous. Unhuman.

Her window had one brass rail running at waist level down its full length. I t reminded her of places like the John Hancock Center in Chicago with its luxury condos. Her own view was comparable. Across a wide boulevard stood more Superblocs identical to her own. Their flat roofs were at her own level, so she concluded that she was on the top floor. These monsters stood in dignified rows across a flat plain for miles, connected by roads. There were some gaps which were wooded parks and the street below was lined with palms. It had the same nice green lawns as found in any good resort on Terra. Like a xerox copy of any city in the Sun Belt. Effects were mixed. "Touchy feelies again. Eh?" How did they know of this? Can They read our magazines? They must be into the self same social obsessions we are.." came another voice in my mind. Sentiment sounded valid.

The other Superblocs had walls of tinted glass with narrow horizontal bands of white concrete. These were more solid and cleaner than our own. Had seen conurbations before, but this one was gigantic and spread out over an endless plain. What deeply got to her was its ominous regimentation. In my mind, I called it Fear City. Urbis Phobia in the State of Paranoia. Maybe on Hate Street? Cute. "Never mind. We may get into that later on." Voices. Alice could not find the horizon which was lost in haze many kilometers away. Were any hills or water behind it? Out of this rose towers which appeared to be twice as tall. Some were thin; even had slits in them, for beauty and to lighten the upper floors. Later on she was to find that the tallest tower on this planet was only 1,400 feet. That reminded her of Sears Tower in Chicago, which held the world record then. Even on this strange world, there are limits to stupidity. And growth. Sic transit Gloria.

There were figures and cars moving down below. Not many but they reminded her enough of Terra to relax her Id. Later on that day, she got up out of another nap. She was at alert at last; had lost some nausea and anxiety of her first day in this suite. It was still July 25, 1978. By only exploring her suite, exercise, napping and pausing to think, Alice had stayed active. By now her muscles were totally slack. Heat had made her strip down to total nudity. Her attitude toward any aliens she expected to meet was this: her looks, speech and behavior would count for nothing. Never mind Star Wars and other concepts, Alice figured. Her Terran origins mattered far more out here... then she had a real shock.

Over by that central pillar stood some new objects: standard office desk, several wooden crates and a nice modern chair of chrome and black leather. These crates were open and full of books, music on magazines and other Terran stuff. On her desk sat one

PC with the famous Apple logo and "64K" on its left side. Stylewise, it was the universal creamy white of the Seventies. On a very **deep subconscious level,** Ms. Roebuck found that her own era was so influenced by that famous movie she had seen in 1968, that one by Stanley Kubrick ad nausem. Now, seeing this familiar Bauhaus stuff somehow shipped or made here gave her the same spooky feeling. Yet it was ignorable. The real reason for that spookiness lay in the premise of said film ... that last part wherein Dave Bowman returns as agent for aliens to dramatically change Terran society. But good or ill? That was the million dollar question. She felt deep inside that Mankind had reached His peak of perfection at that point in time... and that decay was to come soon after that.

In fact, history seemed to prove this. Great events were going on from 1968 to 1973. The Apollo missions and the worst events of Vietnam coincided with that time period. There was even a boom in creating towers taller than the Empire State... but Alice suspected that Mankind would never return to the Moon nor reach any other place in person. Why? Aliens would stop us. Our race in general was to be wiped out; but the Individual was a different story. Woolgathering again. Also books. We had manuals by Microsoft and IBM to help her learn this new skill plus what Europeans call les romans. Scientific romances, perhaps? Yes, these were real books, not just covers with blank pages underneath in English. Like in Earth cities. Yet she still felt as if she were in that Star Gate Hotel in... That Movie. 2001 by Kubrik. But you have to in Your Actual Reality Experience... what Jimi did. Like as they said in that C.B.G.B. Rock Song as it says "Up, Down, Sideway and so all around..." As in, like, cocahina shall get us Way Up There, whereas Big "H" as Horse shall as in chemistry, shall git ya Down, in fact, your organic heart will cease to Beat, as in Your Life Will Expire! Yet if we drop LSD Sandoz 25, say, or any other psychedelic? Then we go Sideways. And so Alice Roanoke tended to think. On any planet.

So what's with NASA? Straight arrows as of 1958. No problem. Too mundane for us Humanities Grads. Ivy League as well. Just no True Inspiration. We need to see God. So what did our hero see? She of the proverbial good movie? She as checked out her books as fast as possible. There were about 300 in all, mostly used. Some were classics like Dumas or Dickens. Yet most were modern paperbacks. Very worn as if her captors had raided junk shops on Earth to keep her busy. Prisoners, they say, need to structure time. No kidding, she said to herself. As if I don't know.

She looked for her favorite reads. Such as 2001 itself or things like Mailer, Camus, Fleming, Ken Kesey and both Vonneguts. It was only Pop Lit. All in English, which was the only language she spoke anyway. Only 30% seemed to be science fiction. The rest were famous works of Terran literature. Most of it was junk. Every day seems to be garbage day in here. This fact led her to think that much of our fiction must be thinly disguised Journalism. That suited her. Why not? There was an audible pop in her left ear.

Then a tiny turquoise dot of light appeared in front of her left eye for microseconds. Her head cleared at once; then she suddenly felt better. She stood up to survey this scene. It reminded her of afterglow of good LSD. Like real shitty thing to say. You think so? Are we responsible?

She picked up one magazine at random. This was a worn copy from 1975. TIME, with Gerald Ford on its cover. Stolen at random to make her feel at home. No illusions created in here. This was not intended to look like any real suite from back home; this is only what average dwellings on this world tend to look like. Now it had been supplied with real Earth style objects. Thinking of food, she dropped her TIME and wandered into kitchen. Again. Do we eat to live, or live to eat? That is my question. Hamlet? This time she made fish and chips. Fried food was best. As was her habit, she sat on the floor while munching and left dishes scattered there. A joke occurred. "Gloria Gainor" was her nickname. As in gains pounds. They got that from hearing about Vanderbilt Junior. Ha. Ha. The aliens were silent.

[**Your kindly put, Editor:** Constant references to "Terra" in place of "Earth" and "Race" in place of vague terms like "Species" or similar gunk are signs of blatant Golden Age Infatuation. Just dig it or trash it. It was groovy from 1940 to 1960 then you just had to wing it with Beatnik stuff and so on. It got more and more Hip to be Hip. We had to take outdated Space Opera and rework it with blatant sexual deviance and drug abuse to dig far deeper under ancient Art Forms.

We need to become valid. This is not "creative" it is Journalism. The avoidance of such silly and outdated terms such as "UFO" and so on is very much intentional. So is an certain gross lack of gender identity. Do not worry, you will catch on.]

[After 15 minutes: Well, do continue!]

Alice was lying on the floor **naked.** Sweat poured from her. **Her Astral Self** got up to tower above her **organic** body. It could **see** a form lying on its side with mouth wide open in deep sleep. For the first time, Alice viewed her actual body from outside. **She was a full six feet tall, or 180 cm.** Hair thick and curly. Somewhat Auburn. She had always worn it past her wide shoulders. She usually had it parted down the middle and puffy, but now it was draped over the floor. She also had very wide hips and a narrow waist. Her breasts were normal for her size and pendulous. Her skin was pale. She tended to suffer from sunburn, making her poorly adapted to this climate. It did not always rain. Basically she was not large for a Celtic Terran, even female. Had good muscles and some fat, but not chubby. This was an ideal body for the infantry. In fact, Alice had been suspecting from the very start that she was being **drafted by some alien military.** Her astral body suddenly left, only to arrive on the ground far below, in front of the main entrance. Suite 6610 was on the North side, giving her view towards the Great Central Gulf 30 klicks away. She saw a huge doorway with many glass doors leading into a gigantic lobby of marble and other luxury materials. Yet some glass was broken, even had been boarded up.

There was no sign of activity, so again Alice had the feeling that **Superbloc 334** was empty. Yet this nice garden was still well kept. It was a typical one for the tropics. The plants in it resembled those of Terra and some were even direct imports. She recognized date, banana, papaya, palmetto, orchid, citrus, and cassava within seconds. Her astral energy moved towards Bloc to have a look at it. Could see through the structure, as the lobby opened up to both sides. Epsilon Eridani shone directly overhead, which meant this place was on the Equator. Astral Alice was unable to float into the lobby. Something was stopping her. **Yet she had no mass.** Here is her report verbatim: "So this must be Seaside Boulevard. Later found it on my PC. Yahoo Maps. There were no street addresses here, just Superbloc numbers. The smog had increased. It was hot and stifling down here. Trees and towers vanished into grey miasma. This boulevard had twelve lanes divided by a wide median. That strip had two rows of evenly spaced coconut palms over 50 feet tall with grass like golf course standard. Stood on a wide path of concrete running from street to the doors looking signs of traffic. There was none. No traffic for many minutes. However, I did hear some faroff music and engine noises, which was spooky. This was not any Terran city, as They call us."

"Suddenly, right in front of me palm fronds parted to let one small figure approach me. Female. Her skin was light blue and hair dark indigo. Four feet tall; age maybe eleven or prepubescent. No boobs. Dressed in nothing but one loose miniskirt of cotton in floral pattern and sandals. Topless. Hair done in typical Sumerian style; but that is also common on Earth. Had gold watch on left wrist. Her features were human, even babyish to suit her age. Yet her eyes gave away alien genes as soon as we made contact. You see, instead of what we have, a white orb with iris and pupil, this being had only pure shiny black eyeballs. Two of them. Not like us at all. Like a crab. This shocked me. Frozen in shock, she stared back at me, then sneered and pointed at my clunky cowboy boots. My clothing was too sturdy for her. Mind you, this part must seem strange but she could perceive my astral body plus whatever clothing I usually wear. Most OBE "bodies" are nude or in jammies. I must have been projecting such visions to this area without effort.

My Dodge with 38 Smith Wesson had been left behind by whatever cosmic forces had stranded me here. **Combined with my size and predatory instincts, both objects would be lethal weapons here.** Why do I have to raise such issues? As far as I can see, this was not Dystopia. Maybe even some kind of Paradise? After a while my forehead grew chilly. **A small blue dot** came into my field of vision. The word Eridan was now audible. Then, knowing I was on the right track, pointed at that big intense fireball up above and heard the words Epsilon Eridani as reply. **Was not vocalized but done by telepathy.** She then pointed to my Lobby, then invited me in with a wave. She resided and worked here as concierge and appeared to be the only person left. This being seemed to be de facto owner of Superbloc 334. Saw a few others like her walking around in the distance, casually enjoying the parks. All were females with blue skin and large breasts. They had the same Sumerian hairstyle along with flimsy clothes. In one group of three, one wore a dress of Hawaiian pattern while the other two were topless with very revealing white miniskirts. Finally, some cars went by. It closely resembled any common make found on Terra, which improved my mood. Alice typed, "I've read cheap novels about the idea of future dystopias. They assume that this soi disant Computer Revolution off this decade will destroy our moral values. Well, I do not buy that theory at all. Even if I was very much against Social Darwinism. My long hair has nothing to do with it. Yet was to find out later that the Eridans were more advanced technologically than we were. Despite huge areas of vacant housing, their society still had activity of some kind. One knew people and cars existed out there. By the way, this area was close to both water and its core. Only five miles away loomed one cluster which had to be their CBD. Central Finance District. Her name was **Et Tiana.**

Etiana just reminded me of Eloi from the Time Machine novel by H.G. Wells. Learned soon about ESP from her because she simply beamed out mental "units" of meaning without verbal content, which amazed me. She told me that she was aware of my conceited opinions, ie. that thing about Eloi. In fact, Etiana was only a **serf,** and not the smartest nor most powerful in social status. Her leaders were of the same race, yet so superior that they did not even appear in their Media. After asking where I fitted in, Etiana claimed that I was not as wise nor as sane as I figured I was. One other detail I learned. They breathed oxygen, but their blood was based on copper, which gave them their blueish tint. This copper oxide was actually green, but some other chemicals made it appear more on the blue side.

[**Editor:** So. Your amorous act gets in the way of your writing, which does have merit. Maybe as un ancient **roman**. Hmm? Maybe you should drink VSOP brandy along with Screw Loose Toulouse, and otros **idiot savant.** OK. Now grok this: The vast majority of citoyen on Fair Xiotan are oxygen breathers. Carbohydrate Units. OK? Yet, some few hundred Insurgent Commandos with automatic weapons, and even Sarin, do fight here. In some covert war. They be Archaea and racist. Then, in the middle classes, for we do have a class system here, we have many millions of Cyano Forms. As breathers of nitrogen, which can be combined with carbon to produce a chemical reaction with energy. And adds Biomass to their bods. That can even happen down on Fair Terra. For your world has air that is at least 70% made of nitrogen. Right? **So go figure.**]

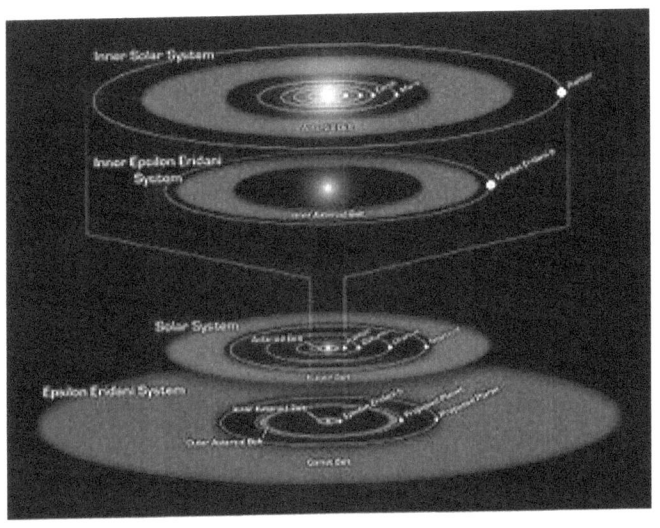

Tried to get more data from Etiana but did not achieve much beyond more cosmic generalities. Alice really did not think this patch of urbanized tropical floodplain was Manaus of the far future; she really did not believe in Bullshit Cosmology, as in time travel backward and forward without solid physics to back it up. What bugged her now was the fact that Etiana wore only her mini which was almost up to her crotch and nothing else besides sandals, golden chain belt restricting her waist and many gold anklets, bracelets and chokers. Her bare breasts made our Terran recall **ancient whores.** Poor being. Banged up any orifice. Alice concentrated on this apparent **slave** in front of her, then mentally beamed, "No Industrial Lights and Magic Bullshit now... who are your leaders? Not you, I say!"

Her vision fogged over with yellow glow. Sounds went out to be replaced by low **buzz.** The scene faded, and seconds later her astral body was back in Suite 6610. Alice got up and stretched. It was obvious that **The Others** had hold on her somehow. By their immense powers over Terran nervous systems. They must be benign in the end; their ultimate purpose being to train her in various Occult Arts. That blue dot coming into her left eye was a sign of alien orders flowing into Right Hemisphere of her brain, which only made sense. So that any nonlinear data being fed in was suitable for that part. The holistic part. One of her profs in Psych 200 had told her that Ornstein was The Book to read. In fact, this theory was extensive and by 1974, very orthodox. He had even suggested that the Right Side was fed data by holographic means, which would then be stored and processed holo graphic too. Left Side of her brain processed data in a binary manner. They flippantly had called it a number cruncher. None of this Body Snatcher activity really worried Alice due to her career as a journalist, her American origins and other factors. The basic reason for her attitude lay in New Age thinking from her generation. It was a total Way Of Life and more or less was said to have begun in 1962 with certain events. (Actually, the real Age of Aquarius will begin in AD 2290 according modern astronomy. But this fact was not known to her.)

Alice knew that she was being totally manipulated by the Others. It was obvious by now that this suite was a golden cage run as school for mercenaries; most likely for whatever they considered to be Real Star Wars. That idea was sporting proposition for any newsgirl. Actually, she was one of those macha chicks who secretly thirsted for conflict. This way, lice had an even better chance for adventure than ever before. How convenient. Above all, here we had the ultimate scoop for her own Free Press: "Earth Girl Makes Good As Mercenary On Faroff Worlds!" She sat down at her workstation to try it out. After some trouble with her DOS prompts/commands, she dove into word processing by typing a report on her first days on this planet. This effort was stopped after a few lines by a rude **C:\\>** then the note: "**Attention.** You from Terra! You can relax and make like one of us! We are not here to kill you! Me only try to be "evil" as in one of your burocrat regimes of Power! They use LSD to melt minds. Was that not evil? Or do you think quite highly of them? What is Good and what is Evil? For the past 10,000 years of copper smelting, and war and peace, Love & Hate, what? Still animal?" The recipient knew at once, that both The Other and she as Alice herself, had read every word of Heinlein's books. And as well, the major works of H. G. Well, and above all, 1984 by Orwell. Must be true believers of Ingsoc. Child of Flavian Socialism. OK?

A:\\> " "Ass!" Alice countered, "Why are you saying this?" Now we have an alien blowhard.

C:\\> "You are not good enough on even your Terran devices! You have only tried a show model of the Apple 64K with Lotus 123 and only once at some trade show last year. Right?"

A:\\> "How do you know about me?"

C:\\> "Call me The Other for now... We have been spying on you for years. Just accept it. When you run into problems with this unit, such as crashing it, then we intervene from our end to fix it. And no data panic. But we still expect you to read manuals by Apple to get it right. Okay?"

A:\\> "Why not zap me with your cosmic grooviness right now?"

Z:\\> "Never mind. We want you to learn your own Thing first, then you may learn alien ways. You will find out why later."

[Pause.]

Alice spent more time in witty but weird repartee with these unseen aliens. Was getting used to their snark. Could have called them "Overlords" after an old novel, but decided to leave that for later. She demanded facts on her location. The time of day for now. **Z:\\>** "Offhand, it must be 7:00 pm now in Terran terms. As in your 24 hour clock. Your PC and watch will both go on like normal. Yet local days have only 20 hours. So, as you must realize, we are not some "gods" who can control Total Reality for you as Universal Constant! Do you understand?"

"Of course, Friend." Who is who and which is which? We are one.

"Thank you, how nice. Now grok this: Our local day is only twenty Terran hours long. As you have seen, our sun rises on your left, or East, and sets 20 hours later by your watch on our Western horizon. We are 500 kilometers North of the Equator."

"What city are we in? What year?"

"This is the capital and largest city of Eridan Two, a planet orbiting our star at 65% of an Astronomical Unit. That ensures Terran comfort zone. Our third planet out, at 1.08 AU is also habitable, but colder. Then come a few gas giants in empty space, until we hit 50 AU. Then comes an Oort Cloud, like your Kuiper Belt, but wider and made of halide gas. You see, planets within range sweep our system clear of junk. This young system is stable and safe." Our worlds [The Other went on] have no tectonic activity and few meteors to collide with. It is hard to observe our system with clarity from outside, which explains failure of Project Ozma. Then, finally, our ancient name for Home is Xiotan.

"I have heard of Project Ozma. In 1960, they declared both your own system and Tau Ceti the two most likely to have life, for various reasons. Now I know why. Can you now tell me more about your city?" Alice typed. "Okay: This is one urban sprawl. Millions of years old. We had 25 million people here long ago, but now only two million are left. Don't ask us why. Hard to explain. This whole planet has been underpopulated since we started some civil war in AD 130. We will call it Gulf City in English, since it lies on the Great Central Gulf?"

"And this planet?"

"Here are some facts: Xiotan is covered by one big ocean and our two Poles are landless. Only 30% of surface is water so land covers most of it. That is scattered across the Equator as seven continents linked by chains of islands. You are on The Great Continent, which has 40% of our total land surface. Five miles north of you are our Financial Center, docklands and then the Gulf, which is 800 miles wide. South of here are giant floodplains covered by thick rainforest. There are scattered patches of farm and small towns. But most of our land mass is selva. As you can see, this entire planet is tropical and humid."

"I can see that. This air is stifling. What's wrong with it?" That was so. Alice was being diplomatic. "We have far more carbon dioxide, half the oxygen of Terra and a pressure of 150% of your own. What makes up for that is low gravity: 90% of Terra. This planet is smaller. We are also 10.8 light years from Sol. Not so far for such beings as me. Too far for you." They were about to hand Alice more data. She made one useless remark, "Me thinks you have been taking me for a ride timewise."

"What do you mean?" asked the Other.

"My watch says July 25. Was abducted on July 20th. I feel that it is impossible to teleport a person all the way to another star so fast. I say you zapped me into a saucer somehow, but it took longer to make the trip. Months at least... you must have tampered with my watch!" Long silence. Alice tried typing but her screen showed no results. Only the silent prompt... C:\\> She could bang away on keys all she wanted, but it was obvious that her PC was controlled by some jerk with its own modem. Well, at least it was not on the ESP level. They were only overriding her on this goofy little machine from Terra; almost a toy. They were expressing much noblesse oblige. For some reason, the Other was being very respectful to her. After her last debate, Alice was no longer worried about paying them back. She now had a theory that they needed her to "save" their society. Another fact by now was obvious, and of great interest to our own science. That big belt of nebulous garbage around Eridan System told her at once why our CETI/Ozma projects had failed. Any "ambient" radio/TV signals coming from some star could never maintain the strength needed to reach Sol System with volume or coherence. Not only do radio waves spread out with distance but huge clouds of "dark matter" between stars absorb... her mind was cut off. A voice from the ceiling intoned **"The objectivity here is 100%!"** It was deep and male, devoid of emotion. Alice was shocked. Then, "We will kill you eventually. A murder. Period. Not rape, beat, torture nor molest you. Just murder you but sans pain... kill you, that's all." That last bit sounded just like the CIA. Which she had no contact with other than Hollywood fantasy. Like most of us. Yet seen in a more positive way, it implied that there was parity. If she met the other side halfway, then she could expect to be treated kindly. Alice looked down at the screen and saw this:

Z:\\> What happen?

"You know what happened." Alice typed in reply, "You got cut off and some weird voice handed me some shit about violence."

"Wait a minute! Yes, it's me. In fact. I made that thing up. Used some Terran male voice we taped. You may call me The Source, as in your ERT movement. We know you attended Reiner Erhart shows. We are not into his head trip, but we assume his Source is what used to be called God on your world. Okay?" Text in MSDOS on her screen went on to say "Since we can control your central nervous system directly by the use of impulses, we have no need for threats. That voice of yours is only one of many found in your cities. It could have been anything... even random ESP from some Superbloc nearby. Understand?"

"Yeah. Sure. Like as in that sappy Heinlein book from 1961. The Old Ones floating around Martian cities. I suppose you'll tell me it's ambient like LA smog or something." God is...? Me too but don't make big deal. "No, Alice. It is probably, now that you mention it, a random message from our enemy. We still have a hostile faction of demonic creatures of our own race but different. In fact, we are making ancient conflict with other Empires from other star systems. Causing civil war down here. They can abuse ESP like we can."

"Are **they** in this bloc?"

"Maybe. We tend to sneak around here hunting each other like in **Yemen.** That kind of thing. Get it?" This Source was beginning to sound dorky again. Bringing up issues. Alice agreed and then explained at length how she planned to prevent more such attacks. She was well versed on that level, and was glad to have a technical toy to play with. This dialog served as welcome distraction from that scary conurbation outside. The Other then gave her some more facts, all of which she could store on "floppies". This whole dialog was storable on 5.5" discs for serious analysis. Among other things, Alice was told this System had other planets, including two giants and was younger than

Sol. The Primary was of the K2 spectral type, had 60% mass of Sol itself and 32% as luminous. Color orange when seen from afar. Then the Others called for a break. Alice had the evening off to speculate.

And so she did. First she checked her fridge for food. It contained most of the standard items found at home. She made a meal of salad, fried pork and instant mashed potatoes. Then washed her dishes and dried them, which was relaxing. Alice craved booze, tobacco and cannabis, but was still tired. A new feeling of having achieved something was the deciding factor. Computer training gave her the same sense of satiation as weapons training did.

As she lay under light blanket on futon they had also supplied, Alice let herself drift towards sleep. A full achievement, personal contact with a rather drippy alien and the security of a good home allowed her to be this way. She concluded intuitively that the trip had taken a day or so on a UFO capable of teleporting itself from Sol to any nearby star. Any extra time was used up by calculation of dissonant Frames of Reference which proved out Einstein's Unified Field Theory. Her watch had merely been left on her wrist where it belonged. Spooky Time Space Paradoxes could have been implied, but not this time. It added up. Then she fell into deep slumber.

[Pause.]

Dream sequence: It had actually happened to her once in 1974. A rainy afternoon in May. Alice still in High School. She had been downtown looking for summer jobs. She could recall herself then dressed in a nice outfit: black nylon pantsuit with matching jacket, wide fedora of black nylon as well. Over this outfit she wore a salmon trench coat with big buttons. Synthetic fabric was in then. As in a pop song, she looked vain. That song going "You're so vain, you probably think this song is about you... don't you? Some underworld spy..." etc. That silly song also had lines about clouds in "his" coffee. Some conceited "man" of our singer who imagined galaxies turning in his java. Heavy, eh? This dream had no humor. Aliens like Other had no humor. Dull, smoggy day got to her. Above all the chill. It got into her bones. She was also tired, having spent lots of time covering Watergate... as a topic for Social Studies 12.001, Sociology in high school counting as an easy credit for College.

She inwardly tended to lean towards the Left. Not that she was against the Republicans in matters of policy per se but she did not approve of corruption. Yet again, she was almost Hawk, having served in the Ohio National Guard for some Basic Training of her own free will. Therefore in her confusion, she represented her nation in micro. Throughout 1974 Alice was depressed, often tired and in poor health. Only thing Alice was happy with, were the reporting skills of the Washington Post staff. She had already tried the Cleveland papers for a job. There were none; she had been bluntly told to get her degree first. That day she had been interviewed in a factory in some lousy area. She went to a phone booth to call another employer. Sorta some half ass Liberal who fucks with Reaganites. and also, this **dialoging** as in how Haig, as in **Al Haig,** used to, like, Dialog, with those Commies? According to that ancient point of view. For money. Yet, the Moon, aka Luna, aka Diana, belongs to us. All of Terra. All four Great Races. If you need to make an ass. That was some LSD kind of position. Give me some groovy $5 tab on blotter. Then me can be Man. Zeno, I wanted to say, LSD is a wimp against your poison of your totally alien body. What of your mind?

[**Well, *** & green onions!** Why not have a greenhouse up there that can hold up to temperature diffs of **400 Celsius?** Glass of some dome can be cooled by its own inherent regulation. Plants growing in earth down below exude water vapor, which condenses on the glass. Which is superhot. Over 200 Celsius at the Noon of Lunar Day. Every 14 Earth days. That will stop glass from cracking up to cause vacuum. It keeps Biome in equilibrium which is natural. Plant our own flora. Then let nature take Her course. Over the seven day Lunar Night, we get down to 200 minus Celsius. That we can moderate by turning on lamps powered by solar batteries. Or even nukes. Such can be regulated by computer progs. Easy. That was what this be for. Me be The Other. You be Man! **Now take what you want!**]

As our hero Ms. Roanoke dialed, she felt a headache come on. As she spoke this got worse until she saw nothing but an orange blur. Her vision was gone. "Miss... Miss...?" went the speaker end of her horn until she hung up in pain. Alice stumbled out of this booth onto wet, littered pavement. Then she simply felt like ditching her high heeled shoes as she staggered about. From far above, a voice intoned "We are called The Leprechaun. Wee Folk. Fairies. Behold in time our ancient and Eternal Enemy, The Men In Black!"

After a while Ms. Roanoke turned to take a look. She saw that a man was now in the booth, also dialing. It occurred to her that she herself must look like the legendary Men In Black herself, in this outfit and in such gloomy ambience. By contrast, this man in the booth had an ordinary appearance. Then he gave her a surprise... by disappearing into thin air. No noise, no flash of light. Alice went to the booth and saw a scrap of paper on the phone which the he had left. There were phone numbers on it. Alice tried it, finding out that it was that of some obscure local firm. Then the entire scene faded into nothing.

[**Pause.** Hot and heavy. Sounds like Brubaker. You know?]

Once more on PC. It was now, according to this machine itself July 28 of 1978 at 2:00 am. The Other was there, in seclusion, typing to her. "Welcome, Terran friend. I am not Your Great Leader. Yet are all united as members of Zenobian Nation, but also as Pantheism! Such are galactic factions. Forgive my hype, Ms. Roebuck but we are just as dedicated to our cause as she is. So what is your problem?" Alice told this new Being about her dream and its heavy implications. After some time, her new Others typed out this shocking message. C:\\> "The phone booth you were using back in '74, which was in days when we had conflicts with the Grey Guys, was rigged from the beginning to serve as teleport chamber. Got that?"

"No, really! I do not dig. What are you saying?"

"That booth was placed there in 1964 by the Bell Company as regular service. But we have records here showing that it was soon after tampered with by other aliens... Greys from Pleiades, Lizards from Tau Ceti, other Blues like us, and so on. They were probably Monotheists, who have been our enemy since AD 130."

"Thank you for the advice. But why a teleporting booth?"

"It is an old weapon of terror. The Monotheists, now under Ariana, selected Ohio in 1965 as a place to terrorize. They used to do the same here, but of course we can defend ourselves due to technology. What happens is this: Some person is given a number at random, such as work placement by Ministry of Vocation, then uses some rigged booth to dial it. This number will then activate a device in his booth, which teleports this person out into vacuum, or onto Luna. Or maybe an ocean." Alice gasped. This news was truly sick. Incredible! And scary as Hell. She asked how such things were possible.

"Our technology is millennia ahead of 1978. The main idea is to harness the organic brain itself, having potential for such action, and then to amplify and direct it. Well, you could read about it in Theosophy and call it the Union of Science and Religion or something, but never mind. It is too complex for you now at this point, so we shall forget an explanation." Alice saw some Eridan faces in her mind's eye. They were those of the minor Blues around her. Etiana, concierge of Yo Bloc, along with some others in a row presumably neighbors from her vicinity. She must be in ESP contact with several natives of Gulf City plus her actual captors. Most of them must be in the same building. Then they all melded into one face: Bold, fierce and larger. Probably their leader. Projecting phantoms. Alice asked more questions. This underling on her nite shift made honest effort to answer them without, A) blowing their concept of secrecy in times of war, and, B) having to explain advanced concepts of physics common to a superhuman society. That would have been waste of time. The **2IC** as Alice called her, which was army slang for Second In Command, went on to say that Gulf City itself, with as others on Xiotan were subject to the same problem... in fact on both sides. Civil war was going on. Alice caught on quickly. "You mean, in this city if some poor suck looks for something, then he can be teleported into...oblivion?"

Death. She typed, "Or if on Earth we do the same, we can run into your Killer Booths and get nailed like this?"

"Yes. Thank you for Too Eye See... how cute! It sounds Orwellian. But as you say, yes we do have this pogrom running on Terra and here. Abduction. Both sides in this local war do the same. You were lucky to have been detected as prime candidate for mercenary and inducted into our Space Marines, hopefully. But again, as of 1947 we aliens have been doing this Killer Phone bit plus other stuff to Terrans at random. Not nice but must be."

"There have been some strange missing person cases in my past, but not many. Thank you for explaining it." 2IC went on to say, "Not by great numbers. Besides, the murder rate in Cleveland has been astounding. You are almost safer here. Down here we have no competition for jobs or housing or other such contesting survival. But the main point is...?"

"I already have it! We cannot go out there into an alien city with some truncated war going on yet. Not by any normal means. I need to learn magic first."

"Exactly... so get with it. This is basically the end result; reductio adsurdem of what your America in the past decade has been trying to be. You will see. For now, relax and dig it... but it is getting late. Please let me go and talk to our Leader about this stuff later on today! Over and out." Then **Z:\\>** went silent. Alice waited for more but it never came. Then her PC simply lost power in some override. Her screen went blank. She went to bed at last and slept like an infant. Why not let ESP as Mommy take over?

END OF CHAPTER ONE

Next: Chapter Two of Book One

"Memories Of Home"

As Alice Roebuck woke up on her fifth day, she had calmed down enough to assess her general situation. Life itself, as Bob Dylan said, was Life Itself Only. Her mind was often taken up by semantics, which is what Journalism can do to you. That PC with the little apple on the lower left corner on ivory plastic for a corporate logo was to be her new typewriter. Good switch from my old Smith Corona she thought. She rolled out of the matrass and sat up on the platform, which was easy to do on such a soft surface. Lucky we have no roaches, she also thought. Any place with wall to wall carpeting that turns into a slum soon stinks to hell, because the critters can hide under the rug by the thousand. Beer, pop and urine soon soak the rug and the end result is one big rotting mass - yech! Reminded her of some dives in Cleveland. Even in its many suburbs. Urban blight that was infamous, for such a relatively small place. Compared to Chicago for example. Then, her mind turned to statistics, truisms, stereotypes and other boring mental traps. Her vocation crept up on her.

The Midwest and how "boring" it was... old stereotype. Well, in the early 1800s it had been an active frontier complete with Wild Indians, bison and... never mind! An honest mind never generalizes unless it needs to see the **Big Picture,** they say. Also, it spoils your fun, as a person to "intellectualize" reality and to "complicate" your life. Or so they say. Those old sayings came from that Thing she used to call **The Counterculture.** But here on a world so far away from her own star system, this creature called Ms. Roebuck was only one more person without any social system to fall back on, at least for emotional stability. But were that actually needed? She was not some Robinson Crusoe. That was obvious from the start. Alice had grasped the concept behind that old book written circa 1750, just as the founding fathers of her own nation were seeking some deep philosophy to live by and it must have influenced them on some subliminal level. Such as Noble Savages? There must have been many of us savages up here as slaves.

The idea was to prove that our modern society was not necessary for survival, even happiness for the Individual. Mr. Crusoe could have easily lived out his life on that vacant island of his, with or without his good little slave who was only a Token Darkie, and that was so dull. If only we could be Pee Cee in Ohio. **Nice colonial fantasy. No imagination at all.**

Alice suddenly noticed that at least two hours of wool gathering had gone by, and without any **voices.** She killed the usual idea of wondering about **schizophrenia** and so forth by recalling just why the concept of Robinson Crusoe and that 18th Century novel had come up at all. Well, in answer to your query, here goes: that guy on his island had nothing to worry about in strict terms of survival. She on the other hand was not some jerk given to romanticizing savages of the jungle; on the contrary, **Alice on this**

strange world had to be the savage. She was not some guy like Swift or Hobbes or Voltaire, who wanted to complain about his lousy society and so forth. In fact, she had always held the view privately that, while the poor of Paris in 1789 may have had a reason to rebel, the Yankees of 1776 really had not. Seen objectively, that is. Then she came out of her **reverie.** The day had by now advanced two hours, judging by the progress of her new sun. It had climbed upwards and to the right from kitchen then over to herself at the bedside end; to about a third of an arc yet often hours. Ergo still morning. She also, once again noticed how bright Sun was. This was definitely another star, as she could tell by constellations at night. Yet this harsh blob of energy up there was the same pure white as the unfiltered light of Old Sol back home. Soon some alien Thing would come in and let her know just how far away this planet must be to adjust for such difference in "luminosity".

At that point she got up and stretched her muscles. Then she ran for bathroom to urinate and wash her greasy face. Later on, our Terran slowly wandered to the window and then got down in an alert but relaxed posture: on her front, raised on elbows. Head up high and gazing out of her window. Without that copper tint and thickness of the glass, that sky out there would have been a clear blue for this morn was cloudless. Temperature 35 Celsius at least, since there was no coolant system. And again, by this star, it had to be close to their Equator.

Alice stared at the most distant towers. They were off to the left miles away. Their main node downtown. Maybe 150 stories tall. Weird shapes, even some with slots in them. Normally they were hard to see for smog. Then she panned to her right. Some nearby towers were as tall. Then came more residential Superblocs. Her own suite seemed to be on the 60th. Reverie turned to abstractions. "In this place" she said to herself, "Noble Savage would be I...." How romantic for Them. There would be no Man Friday for her. No, now she herself was to be slave or pet. She assumed that it was more than one week after her "abduction" on July 20 and now. That interval had been nothing. Void. Could have been a coma of years, yet her body had not changed. Then one more concept. Was not her own Girl Friday Etiana? **No! She was wrong.**

The famous Karen Quinlan came to mind. Yet Alice killed that idea as well. How to react to her jailers if and when they finally appeared in person? There had been a certain long and fruitless debate in some circles, such as in Clarke books, about the "look" of an "alien". That last part of the 1953 novel Childhood's End wherein one of the Terrans gets a tour of Karellen's home world... how corny! Okay, neat trick of Englit: make benign aliens look "demonic". But corny. So what was She afraid of? The sort of life Mankind might have to endure on an advanced alien world was not a casual matter. They may not be into physical torture but how about the mental kind? The Great Society out there was unknown to her and would go by nonhuman ways. Alice ended up sitting at her console. By now it was obvious that there were some Things in control of this situation. Her desk was cluttered with paper and such, like any office at home. They had left her with various notepads, pens and pencils as well as discettes for the PC itself. She started off the smart way: which meant asking the Right Question first... before Making An Ass! Her prompt was the usual **"C:\\>"** and after that she typed "What am I talking to besides just an Apple Operating System?" Answer, "A real actual organic brain, which has the same 1450 cc of stuff as you do. I am one mind in one body. Next?"

"What is your IQ level?" Alice then asked.

"In your terms, 160... and oh yeah, to anticipate your next naive... uhh... **No, turn that thing off**! Vocal Command Modem, you dork! Now I can slow down to talk nicer to this Terran. So Miss...? **Z:\\> Z:\\> Z:\\>**"

"Excuse, man. Why your Americanisms? May be Yankee myself, but why the farmer style? Where did you..." **C:\\> C:\\> C:\\>** **"I am the Other.** Please do not fuck me around. Never mind semantics. We can tell you all of that soon enough. You see, no matter how advanced, no mind of any kind can follow an immature, idle dialog. With any coherence. You are too disruptive. Anyway, we know about your Standard Binet Test and we sort of use it because we have to. Just to give you Terrans some idea of our my level. But you see, that has nothing to do with our collectivism, my slaves, nor our state of technology, etcetera. Like - umm - Alice? Ms. Roanoke of Ohio?"

"Go ahead" she typed passively. This session was better than the last one. More animated.

"Next topic: the nerd who turned on my VCM fudged it up. She is another Organic sitting here with me.

Yes, here in another room but on closed circuit. We are soldiers from **off world** working in part of Penthouse Level in your building. We can use vocal commands for computers but rarely do so. When dealing with aliens like you, better to use ancient keyboard. Avoids GIGO. We do not need cyborgs here." Alice let out roar of uninhibited mirth. She could not agree more. The Other typed on. "We will avoid stuff like that Buddha in my Watch and so on. Such metaphysical debates will be hep in a few years as this New Age fad of yours gets big. We have a lot of stuff that you could not really explain, since you have only an Arts degree. Nor can you build this stuff or personally operate it. However, we can as advanced aliens still use simple tools. Like, if a caveman once made himself a fire, what is wrong with you or I striking a match to smoke pipe? Or light a room in a power outage? Eh?"

"I agree. I may be a pest but when I get a good attitude I can be sweet. But tell me; what is an Organic? I notice you left lower case on in Fonts."

"Good question, Ms. Roebuck. We call ourselves Organics due to the fact that our bodies are hominid, the same size as your own and we have the same environmental parameters. We are, however nothing like you at all. We are Archaean. Our gene pool is simpler and look a bit different. This whole planet is tropical, by the way. But my real point is that we Eridan have other ways of being besides walking around in bodies." Instead of making flippant statements, Alice decided on a good question. "Am I in Epsilon Eridani?" she asked. "Yes. We have told you already. But, as you are in some weird mental state not common to urbanized Americans, you seem to forget. What is your problem?"

"Lousy in astronomy. Just a reporter, okay?"

"Well, you had better learn fast! We may need one genius of an astronomer ASAP."

"Here is more data: This planet is 0.65 AU from its star. Closer than Terra to Sol. So, while Epsilon Eridani is much less luminous than Sol. Seen at a great distance, it appears as small orange disc rather than a larger yellow one. Yet the effect to us all down here, being that much closer, is the same. That ball of light up there is pure white and will blind you, if stared at for long. Just like your own Sun. Our atmosphere on Xiotan, our planet's name, has more CO_2 than yours, so that our pronounced greenhouse effect makes it totally tropical. You'll like it. We suggest you really get into the data in your other files, so you can study this system in detail."

There were some more facts. Alice was pleased to learn that the aliens she was dealing with were natives of this world, Xiotan. Eridan Two in English. Briefly, The Other did say more about her Home World. This star system was young. Only 1.5 Billion years. Archaea evolved here, had created some kind of primitive culture. About 200 million years ago another Race, very different in biology, had contacted them. They suddenly had ships and modern weapons. Far out, eh? Yet to study Xiotan history, and its Native Race in detail, they told Alice to use her PC to contact their local library, to raid their database. That alone would consume huge amounts of time. Anyway, she was trapped in here. Why waste time? It was called the

Internet even here. "Our session is about over. I realize this sounds like those tedious parts in Childhoods End where that UN Jerk keeps asking his alien master what they look like, but we want to keep you in the dark longer. Satanism my ass. So in 1950 no way could some jerk scifi author offend the public. No way, Jose. Not about religion or maybe skin color. Right? Well fuck you. We have certain social problems here that made us capture you as a Terran with no personal reasons and we need to keep secrets. Like this keyboard stuff and our closed circuits. We have sort of Watergate situation as well. Anyway, it is now Noon and we must say goodbye until tomorrow. Over. Was that about merde? As they say in Montreal. I did see that Paris of North America. I am Zenobia Galaxa. In my First Life just according to Karma, was born physically in a body on Io, moon of Jupiter. That was 10,000 Terran Years ago. Then after I did spend many lives like, reincarnated, as one soul did in fact get reborn. Even on other worlds. Like some Tibetan roshi. OK? So, last time it was on Mars. A bit later on, your American Space Program was in swing. That was my present life. So I returned to my Home World Xiotan with my vast Space Army. I wanted to fight Empress Arian to save this world. Zeno went on, explaining Her wars.

"What are you hiding?" asked Alice unimpressed. This Zeno was either a wellmeaning dupe or some cosmic bullshit artist. So what about it, pal? Amigo? The answer was to induct Alice into one big rogue's gallery, as it were. Even beyond what such geniuses as Wells, like Clarke, or Asimov might say. No mere sentimental journey. No sob sisters. Now listen to me: Behold one of our primitive Cyborg like models: Like as if we cut your puny Terran brain from your primo Cro Magnon bod, and like, mounted as some fucking joke into your Crystler Shitmobile aka 1958 Jimi Dean... eh?

"Asshole! Cunt. Well? Out with it!"

"Now dig this: Many times millions of years ago some of us as That Other, as Nation, as Citoyen, we did kidnap and abuse some of you as Beings. Individuals. For science. Curio value. As animals. As mere Entertainment for our children. So there. Like Disneyland in your own shitty Los Angeles. You! Repent! I am your god!"

"What ya gonna do about it?"

"As a whim, a mere folly. My expert surgeons can take your shit for brains out of your body. Like them Infernal Bumhead Mammy Jammers did. In flaps. In 1965. In 1976. Scary, huh? So dig this: We can take your brain of a standard Hominid, like 1430cc. Then we can shove that into any platform. A Warship. That will do you many favor. Make you Great Warrior. Live forever in your machine body. We can put it into one small wandering robot thing. Small. A slave machine. But you are in your natural body. Consult The Gamesters of Triskelion. Original Series. Grok?"

"Was that a threat?"

"Am I then your enemy?" The Apple went blank. Shutdown. No for to do. Just wait this out. Time and space. After that Alice got nothing but the usual DOS responses which were limited. Even she got wise to the whole scam, it came down to the simple fact that we humans can easily program a limited range of responses to standard commands. Such as if all phrases typed by any human operator, no matter how meaningless, had to be answered by some politely worded phrase...? Well, we are not even dealing with Terran Beings here. Anyway. Nonsense would have impressed her back in high school but not now. She deduced that any Earthian brain could be inserted into some kind of alien machine made of metals, plastic and glass. To be Their Eternal Servant, Slave and Soldat for much longer than anybody could live. That was still dependent upon advanced blood supply with sufficient oxygen. Wellhow does that suit you, Alice Roanoke? American citoyen. Of them who are true beleivers of God and aliens landing. That legendary of Yore 62 Percent. So mote it be.

"It's merde, my friend. I will not do this. I am a warrior. I believe in your own Belief. I can hurt you. I can kill you and degrade your body and brain. Go figure. How about your soul?"

"My soul? I have no idea. You tell me. Are you God?'

"I am a very evil and hostile Thing. Beyond your Terran ways. Me need you as my personal amigo. Ally. I can in hate subject you to obscene abuse. Some small metal platform. With globe of glass on
top containing on human brain. Eyes attached. With mouth to speak with. With batteries, blood supply for one brain in one small mobile pack. It walks, talks, can transmit MS. As in Gary Numan. Organic wedded to machine. Is that good?"

"That was an obscenity! Give me an alternative." The answer came soon. Not some Gothic Horror stuff. Zeno was okay. So Alice had her organic body. So did Zeno. No conflict. Consensus. Alice sat back from her desk with a mug of good coffee which she had gotten from her "advanced alien kitchen" just minutes ago in mental fog. She had used her General Electric coffee maker. Which they "created" here as well. Hmm? Then she noticed that there were more items that aliens must have carried in overnight undetected. Books, record albums and tapes. Boxes of conventional clothing in her own sizes, such as shirts and pants. These must have come from some Ohio discount store. She later found that one big box contained three US Army combat uniforms. They were the usual olive green and came in her size. Over by the door: three pairs of Size Nine US Army combat boots, standard black leather. That suited her best. She had always had a secret wish to be a war correspondent.

[**Editor's note:** What? Rough, eh?]

After some time she realized that 80% of these items were stolen from Terra. All the cardboard boxes had common US brand names on them. Yet the rest of her things were copies of what Terrans had used for millennia. That last fact made her think, as it implied that aliens had to pander to other prisoners who might be here as well. That reminded her of the way Morlocks in her old Time Machine novel had carted in food for the Eloi... and then dragged sleeping bodies out to eat them. Ugh. That bothered her. "Frankly, no." she thought, "They won't eat me. But this sneaky business bugs me. She finished coffee and returned to bed later that afternoon. Bored, Alice drifted off to sleep. Just before that she noticed how the shadows from some of the taller towers had been cast across the floor, as if someone had placed a few stereo speakers along the length of the window. These shifted as with Epsilon Eridani as it glided along its path. The room was bathed in a warm amber glow. Alice vaguely recalled some tiny figures walking about below along with cars and even aircraft as well, but they had not registered on her mind as anything.

She remembered long hot summer days in Cleveland just like this. Twas back in 1970 that she had finally grown up as it were at the age of 17. Had mastered her most difficult subjects back in high school by applying some philosophy; namely by dividing her world into "attitude" and "reality". Instead of trying to childishly copy Adult role models, such as Hippies versus Squares, she decided on simply observing daily reality as found in her own social circle. Most of us eventually do this. It would have been okay in some Romantic way to try to rebel against the entire System as so many kids did. However, she did not. Alice had always been a bit spaced out. Not one person had ever used word "schizo" on her. Nor "mental". Ever. Not until 1978. We have no concept of the Why. Yet were such words only that?

In the Sixties and Seventies the very concept of "mental illness" had not been regarded as being any valid social issue. She had many friends and lovers as of 14, then even more friends, sometimes too many for her own good. Okay for many fields but not STEM. Her physical condition, size and stamina almost got her into police or pro sports. Her figure was sexy and looks were almost fashion model quality, yet she had no inclination towards being bimbo in public. No Bardot here. She drifted into college like some kind of amoeba. How useless. That is where her reveries ended. She drifted off into sleep. One smartass note: A rather daring shrink had once suggested to some encounter group that "depth analysis would destroy" them. Her reply: "No, it would not... in fact, it would result in me destroying you!" Said analyst never asked whether her intentions were physical or otherwise. Maybe he was okay on humor.

[**Editor's note:** College humor, eh?]

More digressions into the past, all coming at her in a series of visions. The word Experiential was often heard in New Age circles on Terra and even her captors had voiced it at least once. It alluded to an attitude common to hippies or dopers. The same one Camus and Sartre pushed long ago. The idea was to enjoy pure experiences; not to become obsessed by them or try form an attitude by using them as raw material. Study and passed exams like a robot, without any emotions nor fun. She had, by the way, been intensely interested in the News as pure data in and of itself and was fully capable of displaying her intellectual capacity. She had often, as favor to profs, attended parties in faculty lounge to clarify some point. They accepted her as a real person and found her to be a useful link to her generation. These party gaffes had often helped teachers redesign their lectures to suit the demands of American society in the Seventies. Who was Allison in that Elvis Costello song? Keep this on an intellectual level. Sunset came at 6:00 pm. After that, she tried to get some response from the PC, but ended up with nothing. She turned off all her lights and noticed that there many small lights all over the place at all levels. This city was by no means fully occupied, but there were signs of life in some of the upper floors. The whole thing was lovely; it resembled our Galaxy itself. That experience alone kept her from losing sanity night after lonely night.

The sight of moving lights below reminded Alice of her home town itself. Moving cars and mysterious faces of this Cleveland surrounded by the blazing factories of Ohio. After she had fallen asleep, vivid dreams of her past life flooded her mind. They started with 1978. REM action took over. Then it finally clicked. It was Really like, Happening! She was no Mr. Jones who did Not Know. The Score? Alice was only her frank naked, personal Self. Inner Self. Core Being. Her astral body floated up over her Superblock, leaving her tired, wasted mundane flesh behind. This was only for some time only, not Eternity. OBE. Lights shone and moved below. Far beyond this conurbation, its towers, lay inky black void of The Great Gulf, briny ocean waters. Then, inland, she saw more black void of selvas. She was on Circuit Five. Reminded of one Frank Zappa song, City of tiny lights. Probably based on THC events. Hash cookies, eh? That Frank of ours. Then dreams of Yesteryear.

[Pause.]

Her first dream that night started with one June evening in 1977. This was before she had graduated in her Major in Journalism. Why so picky? Pushy? Oh, so defensive about your Intellect? Alice was intern to some Loser Program as in, work for the summer to pay your tuition, kinda thing. So she had to attend the World Scifi Con in Chicago '77. She was paid for this, but it was yet that old cliche YUMPie Thing of Working As Fun. She liked this genre anyway. In her gonad like way. But needed money. She flew over to Windy City on some cheap domestic "flite" to grok her Con. She was Fen to the max.

Arrived at her destination at 5:00PM on time to present coupon and receive her Prepaid Membership. One whole weekend of free nosh & booze for only $20. There was certain Triple "A" hotel high up in the "AON" Building. Palmyra Inn Hotel, far above crowd. It was not costly nor hard to get into this World Con, no mean Babbit affair but it was of luxury class. Wore her baby blue polyester pantsuit. Carried much ID, her sacred (really?) Press Card, and credit cards. Then she got into her work. Friday from Five to Midnite involved lectures. On books, movies, comics and theories. Fact tended to mix with fiction. "It was only my own fiction, stupid" as Asimov said, but then, as younger Seventies Fen declared, so what the fuck? Extrapolation rules, okay? This is a mundane event. It is based only on science and some magic, but not enough to really notice. Alice went to sleep in her private room. This was in the AON, which is one of the three most visible towers. It is over 100 floors tall putting it in the same category as its close companion, John Hancock, which is slanted and dark, and occupies this turf called Golden Mile. Farther away lies Sears Tower, tallest on Earth back in 1978 but created with a ziggurat design. So Alice fell sound asleep on Friday. By U.S. reckoning.

And as it came to pass, in this wine soaked yet inspired comic opera, that she was to witness, on Saturday Nite, it being Disco Nite "77 in funky town, strange beings from some other star system. Not "Titties n' Beer" by Zappa in 1979 at his zenith of dissolution, but true divine intervention. This allusion is to a common idea that aliens are often just organic creatures who are mistaken for spirits of some kind. She did the usual academic tour of lectures. Then at Eight PM wandered randomly into their Cosmic Dance Party. Deejay played randy tunes. Mostly harmless pop which can be called "disco" but this was all fun. Dancing was in. The Seventies were still alive. The usual middle of the road types were dancing in pastel outfits. Leather and other "violent" gear was not yet popular. More sociology?

Alice danced with herself. It was in character with her Don Juan mindset. Which came from books. As time passed she felt boredom. Then she saw clearly one strange couple. One man was accosted by some Thing who was not of this Earth. People did not react, due to glitter, lights and sound. Drugs. It did not look typical. This being was some officer of Zeno. (A fact not known to Alice at the time.) Was five foot ten. In black teflon weave jumpsuit. Like some pilot or perhaps baggage worker in an airport. Spaceport? She had color of skin like Erzulie. Blue like Atlantic. That color did change, as if flashing in and out of another dimension. Her eyes were crablike. Not in any way human. She asked, in English, if this random Earthman wished to accompany her to hotel swim pool. Why? Just to be water brother. Cosmic "Bee" In. Logic behind that must be according to Heinlein who was outdated by 1957. So he met some beatnik, dropped LSD, then wrote his 1961 Magnum Opus. She, Officer Norom, suddenly vanished into air. Teleport? Split this sorry scene. Bored maybe with this party of mere

This means to say to warriors all, who have braved more evil than Mankind could suffer. They have been pathetic as of 1973, so we must all agree with opinion of our Empress. No. This Race shall never land on Luna again. Such was Her mentation. Not Liberal. Alien. It vanished in not with smoke and noise but just did. Why go with an Earthian mate when this cosmic Banshee can be yours? They read our books, right? That is what we think, Alice later did figure. Caesar and Cleopatra? Was this party ancient or modern? It was good. Some years later, strange folk in black outfits became common enough but Alice and the other dancers would not have reacted to that. Even later on some Pat Benatar wannabe with blue makeup would not have attracted any attention nor even aversion. Not on deep emotion level. Who cares about Youth, Hippie, nor Punk? Just fads. Then again, is The UFO Thing not fad also? She gyrated on dance floor. In ecstasy, but Platonic. Now can anyone tell me that alcohol is not a drug? It's as strong as any other. Cleveland, like any other big city had its freaks and tragedies. Alice had forgotten this fluke event soon after it had happened. Over time, she had heard Strange Tales of aliens and so forth, and even used them for essays. It had helped her career. As student, she had noticed that the UFO history had a pattern to it. Peak year for the USA had been 1973, then caseload got lower. By 1978 very few UFOs were being reported. Once even attended MUFON meeting. Afterward, she interviewed their Regional Director. After hours of intense debate, which resulted from Alice finally asking questions about the main issues found within the field. It was, she was willing to concede, just another valid field of common scientific research like any other. Towards the end, Speaker (whose name is unknown) admitted that the hardest issue was that of whether "aliens" were Matter or Spirit. Well, more or less, in those general terms. As of 1947, starting with their old, symbolic Mount Rainier Case, that had been their main headache. It had also been what had kept both Vallee and Hynek, two Exobiology superstars busy. It was also an issue Alice herself had nothing to do with until she was suddenly kidnapped by this very Thing Itself. Now, on this distant world, she had her chance to solve this very ancient riddle. It was even older than the Sphinx. Are They indeed divine, maybe Gods, or only animals like us?

However, Alice made few meaningful connections in her mind about UFOs. In those days the idea of "War In Heaven" between various groups of aliens was not common, nor were other ideas. Before 1980, most people in the United States did not think about that issue much. As result, Alice just drove out to meet her witnesses and then wrote standard UFO reports among other news. It was a sign of her naivette, among her fellow reporters, that she was "soft" on such bullshit. By 1978, no real Media Man would touch issues that bogus. Like that "man" in the song, she was The Media. Twas her identity. It grew on her. It was as if those three musicians on their island the symbolic and that haunting opening credits from Paradise, aka that ancient and famous show Here Come The Seventies, had failed to spread their message of Universal Love.

Alice woke up in the wee hours of the next day to turn on her PC to carefully record first dream, which was of Blue Teflon pilot being overcome by fear and loathing in a scary alien city with no Fear involved. No Hate. But what of Love? Is Love same as Sex? Me wanna but cannot. Me what? Converse of her own personal situation! Terran as They say. I am now, with my proper English, trapped here with concepts. They would be Mayan concepts such as Nagual, Tonal and Units of Meaning. Such might help in such extreme alien culture. here on Xiotan. Again she drifted into sleep.

"I know you. You do not know me. We can kill you." quoth The Other. And God? I cannot say more. Some voice from darkness. Alice guessed it was ESP from some organic beings somewhere in this vast hulk. Save for her own lit up window, which was visible for miles, this mass of concrete and glass was dark. Almost uninhabited. It was hard to get used to. Alice recalled a panel talk on "Hitech And The Next Decade" as amazing if somewhat gloomy issue. About the coming Micro Chip Revolution. There was large crowd for this one. Alice got to see a portable Notepad and wondered about buying one to use instead of her klunky Smith Corona. The first real word processor was thrill for her. Her Con came after that. More scifi, eh? That was the "fun" part.

Her next flashback (as dream) occurred the same night but on Xiotan. This, my poor readers, despite any Time and Space issues. So Adversary Dejerk awoke at Dawn. (A nice girl's name, eh?) She put her boots on? Then fell asleep again only to dream her Gardnerian mode again. Wet dreams? In fact, though leaving humor aside, in her dream, Alice stood in line at busy counter in some bank in the Grover Building right downtown in Cleveland. Her dream, of course, was about an event that had happened back in reality. It was months after her Con. This episode had started at 4:55 pm on Oct.

22nd, which in the Midwest is always official Daylight Savings. She had not checked any clock yet. Had her card, out teller could access her account but no PIN number for we had no banking machines in those days. Even "client" cards were new. Had to write some long number in addition to her account down on paper, then show her card. Etcetera.

She made a routine deposit, then turned to leave. As she did so there were many people behind her as potential witnesses. At that point, a Thing appeared directly in front of her out the "blue" as it were. The same alien she had apparently run into once. In addition to Her minor officers, as Empress. Aka Time lord. This time It wore another outfit. Instead, we see Natacha the stewardess from Belgian comics: Dark blue jacket over a white linen dress shirt, miniskirt to match the jacket and white shoes below long, sexy legs. She had the same hourglass figure as previous aliens. Only: Blue skin. Bug eyes. This Apparition smiled and asked, "What time is it?"

Startled, Alice looked at her watch and read 03:52pm. She said, "It is about Ten to Four, I think..." Rudely the alien interrupted to say, "It is almost Five, Miss! Which is closing time." As that came out, Alice was still busy staring at her watch and could see only a shadow in front of her. She looked up but saw nothing. This Thing had vanished by magic. Alice walked away in silence. That had happened in October of 1977. Alice could easily recall the incident later that day. She was sitting at her PC at the time. Then a new message flashed. It was from some unknown source. It spelled out in big green letters, "Bravo!"

"Thank you, Other... the Source. Whatever. I really got strong results out of my dreams. They are mine, after all. To cherish and to hold as they say. I feel relaxed now. So now what?"

"Look at this." the Other said as the screen filled with regular TV content. It was as clear as possible. A hand held up one black & white 8 by 10 glossy photo. "Oh fuck!" Alice shouted, "It is my double." It was in fact of herself standing there in a cafeteria at Kent State smoking a joint. She wore her usual green Tshirt with jeans. There was nothing behind her but vacuous grey sky seen through glass, plus tabletop and chair, indicating an upper floor.

"The question is, do you recognize the photo?" asked the Other. "No. The person, yes. It is me. I recall the place and time no problem. We often spent time in that cafeteria on the Tenth Floor back in College. Munching, having beer and grass after study. But never had anyone take photos of me there; certainly not smoking dope." The Other said, "This photo is a real one. Sure, with the hitech we have we can forge such things easily. But in your case we wanted reality. So we had someone come into your cafe and take this photo. It was some local Terran with a telephoto lens. Paid by us. You simply did not notice. Is this not Orwellian?" Alice felt like laughing. The fact that these nasty aliens had been treating her like royalty for days now had relaxed her. This was like Club Med; not being a prisoner at all. So she said, "America, Russia and China back on Earth are very Orwellian. I see no such thing on this planet ay all! Your sneaking around, taking little shots and so on are nothing compared to what Nixon did, okay?"

"So you deduce that we aliens have good reason to maintain secrecy in your society... in doing our tests on you? Why not bring up Brezhnev as well? Is it not degrading to be treated like a lab rat?" Alice burst out into laughter. She was cynical about any kind of Orwellian thinking. "We agree. But please! Why the Newspeak? Why not call Eric Blair Great Thinker as Humanitarian? Instead of two minutes of Hate, why not five? Why so cheap?"

"You are right, Alien! We of 1978 are as debased as them losers in that book, and honestly expect our own future to get worse. Still, it is only a satire." There was a pause. "Again we agree. In general terms. Miss Roebuck, let me say that neither you as person nor your Race is on trial here. This is no rat cage! No! We need to exam you Terrans for the sake of our own Anthropology Section. We need new data. Just to survive. Capiche?"

"We speak of Orwell later. It's a worthy subject. But for now, The Time Machine by H.G. Wells I'd rather get into. Your city reminds me of it. What is its name?" There was long pause. Alice assumed that there were several creatures in some control room in the same structure sitting there debating their methodology. Not scary in the least. They finally typed: "Good idea. This town is called Gulf City, and was once capital of Xiotan, lying on the main continent. The Time Machine must be relevant to us, because we have similar problems. Gulf City is now almost vacant. How can we hide the fact? We are a culture over One Billion years old. The Biome changed from Flourine based to Oxygen based about 80 Million B.C. like, from Archaea to Eukarya. This city, like London in AD 802,001 is giant tropical garden full of empty structures. Used to house 25 million, but now has a mere tenth of that. So we can really identify with Eloi and the Morlocks. We have in fact similar social conditions right now! Yes, like in the book." There was pause. Alice asked them how much like Terrans they were, at least mentally. They said she may as well think of herself as being in the same position as the hero of the Time Traveler, which in their own opinion was the best work of fiction ever written on Terra. So wide was the gap between their cultures. "Okay," the Other typed out, "we can admit to having vocal chords with lingos as on Terra; although we mostly use ESP with each other. Some of our basic concepts of society influenced the language of Sumeria as of 10,000 BC. which in turn created some of your modern... wait! I have gotten orders to come to a stop. Your case can wait. You also need a rest. Until tomorrow at Eight you have time off. One more question!" Alice asked "So about that funny thing... what was it?"

"The photo?"

"That, but also that other thing... was it Objectivism?"

"Ha! Now we have you! You are so reified. Yes, you are. We take your overly academic mindset as the norm for now. This is not about Ayn Rand, Mod Lit nor any of your philosophy. You have an Ego Problem and a cloudy mind. For now. Anyway." The Other went on, "We decided in 1947 upon our return to Terra, with few craft and only as one race of many to be really blunt about our contact methods. We needed soft touch. So we made photos and other secret records of some humans. Only one from the USA. We started with a few others but all were rejected leaving Alice Roebuck as the only good sample of modern Cro Magnon Man. We used only Terran methods of recording. In fact, we try to almost think Terran just to become as Terran!" That explains Narco Action.

She found them coldly kind. The Other went on, "We made some photos and wiretaps. Imitated Watergate. Sorry about the spookiness, but then I guess that says much about your nation, not us! It was only later that one of our scholars pointed out how much a certain scene from the Book 1984 resembled the time we showed you that picture from college. Think, Alice!" Alice smiled. These aliens were into something like "You are OK; I am OK". That suited her. Some pain departed. The Other said **"So here is our real point:** We as culture are more like the decadent Eloi of Wells. Far more than your strident Oceania of 1984. You would be better off in that latter universe. Can you agree?"

"Yes." Alice said, "I know what you mean. Felt more at home in 1984 than in 802,001. You are not my People."

"We said 'The Objectivity here is 100%'. So anyway, that phrase comes from some CIA Handbook. We Eridan spied on your own Central Intelligence Ageny long ago and found out that it was one favorite opening speech handed to every prisoner. Those would be in the Third World. Remember? Then comes stuff like, 'We will not rape you, beat you nor starve you... just kill you that's all". And like that. Assume the East does the same. But what do we mean?"

The Other continued, "We are too sincere. We like you as person. We need aliens with good abilities to help us save our culture from final decay. See us as Eloi. Help us. The emotions we have for you are 100% positive. In other words you are here as our equal even though you were taken here as a prisoner. You cannot leave this planet without our help yet you can easily survive here. So this is vast prison camp from your point of view. We as Keepers have to be objective or else what is the use? So we took concepts used by your leaders. Only in a nicer way. As close as we can get to some benign treatment. To be objective. Now over and out." Evening of August 17, 1978. Thus, Alice decided to indeed be "objective" in response to the advice of her alien friends. They finally came across as such by now. To her relief. She had expected worse. Some corny jokes came to mind.

The great orange ball of Epsilon Eridani was swelling as it set in West, which was towards the right wall of her suite. The shadows flowed across her carpet. She just sat by the glass in total passivity allowing the alien scene to soak into her emotionally. Much the same as she had handled the apparent "culture" of her new "friends". Just go with the flow, she figured. Some breeze blew her hair a bit. There was no air conditioning, but fresh oxygen came in through small vents near the ceiling while CO_2 was sucked out elsewhere. She had finally found the controls for this system. It allowed for volume but not heat nor cold. Air in here was a steady 30 Celsius plus. She had donned a combat uniform. The front of her shirt open as usual to leave her boobs exposed. She was childishly receptive to her alien adventure. After sunset, she fell asleep on the carpet. She quickly drifted off into dreams of her past back on Terra. However, since there was no alien influence on her mind, she only dreamed of her normal life. Her experiences with those Azul beings from 1974 to 1976 were forgotten. Yet one more new day under the blaze of this distant star was coming.

END OF CHAPTER TWO

Next: Chapter Three of Book One

"Gulf City"

Next day Alice woke up in time to do some exercises and start another session on her PC. She began with some comments on "objectivity" as way of life. The Other was decidedly obtuse today, which was strange. They were being diplomatic about something. So to help matters our prisoner from Ohio began with generalities. Once more, she stated that some general concept had made her life worth living in the early 70s. She may have come across as being trendy, vacuous and insipid but it was okay in that decade. Basically, Alice felt that Terra and its 2.5 billion people had made a paradigm shift in their group mind then. She had often met people who were still into "copping an attitude "

"Funny way of putting it." commented the Other, "Don't you really mean they were being Idealists?"

"I get the picture. And I as Cynic or Realist. Whatever. You are playing at semantics."

"No problem. Actually, you simply adapted to changing times to ensure personal survival. Check this out: Your friends must have used the expression Becoming Aware a lot. Awareness. Is it not so?
I do not mean to be corny."

"I agree. We were not corny at all. The whole thing was for serious purpose. Survival. I still had ideals but became more of an animal. More instinctive. Yet also have more intuition and more intense perception." The Other said that was good response. Their project on Terra was partly to find out what "attitudes" Terrans had as time went on. Of course, they had to avoid becoming too flexible. Average person came across as "raving" empiricist" to their own leaders. That got a laugh from Alice, which was ignored. Type went on... C:\\> "What we mean is that we have a certain Philosophy of Life per se, and that our enemies, who exist on this world, always religious factions who dare to oppose us... do not share that with us. We disagree on many fundamentals of Truth!" There followed more bullshit about Religion, then, C:\\> C:\\> "Ha! So you admit it! You jerks are a fucking theocracy! I won't even capitalize it. As you can see, we Terrans are mostly not into that old bullshit. East, West, all over the globe we are getting into secular humanism. I got into it as well. Try the word 'cool' sometime. I'm cool. So you think you can Shanghai me into your crusades? Is this really your galactic empire? Like in my Foundation Trilogy?"

"We really had you going. Didn't we?" Both sides were having fun. "But this is no mere Exercise In Semantics 101. We need you for our faction. Our gods demand it! Like this civil war we have been waging on this planet for 18 Terran centuries. Semantics, ethics, attitude, mindset, concept, mores. It is all the same! We can call it the Philosophy of a race. Whatever has been guiding us for 80 million years. Comments?" Alice needed rest. Just by dropping hints she had scored bigtime. They handed her some real meat within minutes. She excused herself, made hard copy on 5.5 inch floppy diskette. Then she had a meal of coffee and toast. She noticed there were various Terran fruits in her fridge; brought in by elves or Morlocks perhaps? She had bananas, oranges, apples, plums and kiwi. But not any "hypertrophied strawberries" as in that old book. She ate some more then had a trivial idea. (Yes, they crowded into her brain a lot.) Then walked over to Apple peeling banana. What were aliens thinking? Alice sat down and munched some more. Then she typed, "Can you take orders? Deliver stuff? If so, then bring up several potted plants. Not poisonous ones; just decorative ones. Like palms or ficus. This place is so drab."

"We have various houseplants. All are standard Terran types. We can have our slaves bring 'em up. While you sleep. But no local flora and absolutely no fauna. You are still under quarantine for disease. What else?"

"Pizza. Hamburger. Fries? How about furniture?"

"No can do. We have no Terran cuisine as advanced as pizza! But we can make wheat flour, margarine, spices and other stuff. Ground pork for meat. Cat, dog and fish as well from Gulf City Zoo. Also, since your desktop PC corner is getting lonely, we shall bring in a couch, coffee table and more chairs. We also have TVs and VCR stuff, plus some star charts. Are we okay?"

"Thank you. Now how about the old Schule spiel?"

"Good girl! Now on to Philosophy 201. So as we were just saying, the guiding ethics we have are complex and ancient. We follow cosmology similar to that of the Sumerians. In fact, we call it Applied Cosmology. The math came from the Galactic Empire itself and is billions of years old. We shall try some now:

0 1 2 3 4 5 6 7 8 9 0

The above eleven digits are actually the real way of presenting the entire series of numerals found in the Arabic System, the one still used by modern Terrans and also by us. Notice that there are only ten, so that "Zero" must be in fact a nonexistent number. We have been using Zero again as notation for "Ten" because we did not wish to repeat that "Numero Uno". Zero was only Placeholder. What was the reason for that? Alice answered, "No idea. This playful math is not my thing. I Prefer geometry. I give up. What gives?" Was tired. "Okay: the two zeros; as you see, are at both sides. It is correct by various forms of Math, which is an art form in itself. Hence my point. Your visual clue is the same as your reality... thus as you can see, your two zeros visually form Infinity. Even as symbol."

"What? How?"

"Look here: If you were to remove the nine numerals that lie between them, they would collapse into each other and join to form two circles, which resembles your Infinity symbol. You may think (again) of this as mere exercise in semantics but it is not. Beyond any quantifiable amounts; or rather beyond the extremes of lowest and highest possible numbers, lies Infinity. So there!" Alice found this lecture a bit snotty. Were they only telling lies? Was this a movie set with backlit paintings? She asked, "So you are math expert? Suppose you're also into highly advanced physics?"

"We use computers for that. On ships to calculate FTL moves. That Science older Races gave us was not our own. We then apply ourselves to magic, such as our blue girls on Terra. Fairies? Right?" That came from her screen. "I guessed as much. Science invented by some ancient genius far away and not even human. An alien mind." But why not go for it? She brought up Lorentz from memory. Hendrik Antoon Lorentz of early 20th Century Denmark.

"Lorentz, you say? One of your Great Thinkers? You must have attended a Physics Con for pay. Smart girl. Only he was Dutch not Danish. You were thinking of your Copenhagen Con."

"Here is my quote: Lorentz Curves as result of wave mechanics leading to quantum theories leading to FTL. Motto: the microcosm joins macrocosm somewhere in Infinity."

There was pause. The Other finally said, "Amazing! You have no degree in this. We must get into this later. For now keep up the good headwork. We need a break, then we can at last get into your mission here... over and out." Alice sat back and had smoke (both cannabis and tobacco had been provided for her own abuse) just to keep her hands occupied. Slowly blowing it out, she thought long and hard about that ontology session which had just passed. They were basically training her mind in alien ways. As if she was being tested by

Robert Anton Wilson and his freak circus. As if activism of such a goofy kind had any effect on society! Not here nor back in Ohio. She recalled that sad event at her old campus for a moment, then dumped scorched filter into an ashtray.

That was another point that baffled her as she tried to concentrate on abstractions through Fifth Circuit haze and also forget trying to figure out whether her carpet was flammable or not. This stuff about Arabic numerals, electronics and then the old Lorentz Curve. Yet it all added up in her own mind. It slowly crystalized as higher wisdom. Not in philosophy but as engineering applied to conquering the Galaxy. She recalled that back in college, some college friends in Physics had shown around a short essay called the "FTL Paper". It was only a dozen pages long and had one diagram on its last page showing some complex wave pattern. Was it an electron orbiting its nucleus or a ship bombing along out in Space? Alice decided to ask her alien friends via **command.com.** Finis. Next MS on screen: Your Lorentz Transformation not only in like microcosm, was is a time dilation episode... which we do often with our ships. And more. Wait until you see our Quantum Computers. We shall land in Essen, down in Union de Europa. According to plan. Once there, meet one of her mysterious keepers. That was on her PC left one day by some freak called "Zenobia". Alice had never heard of this name before. Who or what could this be? C:\\> More data on DOS. More about it in next PC session.

Alice had not known about any Lorentz Curve nor about his "wave mechanics" nor about "quantum" anything until 1975 and that only because some prof in Physics had finally told them about it. Later on she had even looked it up in the library. The two curves had matched. Then there was the other stuff, all of it hinting at was called Simultaneity... she got up and looked for a suitable book. After a while she found her favorite: The Dispossessed by Ursula K. Leguin. Joke from Other:"Goto WC and recycle DOS."

C:\\> Yet another milestone in her fave genre. As this day wore on, lost herself in yet another dreamland. It was not long before Alice was bored and curious again. But it was better to be curious yellow than bored. Back into dialog on her Apple. "What do you wish to know?" her Other asked. "Which one of you is it?" Alice asked. The answer was more or less the usual; in other words some ghostly entity was there because they were all replaceable like some phantom army from her own Dark Ages. Yes there was even something Islamic about these critters. Amen.

[More Alien Data:]

They explained in a monolog that Alice had to revise her thinking on Physics. Popular concept has it that Einstein in his many theories was "opposed" to the idea of anything just "passing" his Speed of Light, called "C". That general idea seemed to come from Lorentz, who did more detailed work on Wave Mechanics, which later became Quantum Physics. Both had involvement in "quantum" work, so why should there be any conflict? This work reached its peak in the 1920s, was very broad and international, and these people often cooperated. Relativity Theory never said in any rigid way that "C" was some kind of concrete barrier. It was just hard to survive such high velocity as some organism. Such as Alice herself.

"What we do is Applied Cosmology" said Other finally. "That means frankly, that my Race long ago stole ships and data on many fields of ancient science from some other Race. We are warriors, not traders. It was easy for us." Then her screen went blank. "And here we are!" was the reaction. Alice relaxed. Yet another piece of this great puzzle solved; it was balm. She was joining an advanced culture in a friendly way. Then her head hurt. Migraines? She heard a voice whisper into her left ear, "Instead of presenting an attitude... why not really stick up for yourself?" It went on like that. An old and wise mentality, like Obi Wan Kenobi. That struck her as smarmy and cheesy. Almost as if Hollywood was valid on the high academic level that her own story implied. Ennui of now having such things fed into her brain floored her. Time for headache pills, coffee or sleep. She ran for the WC.

[Pause.]

One time Other handed Alice some long speech on "Terran" history from 1492 to 1985. It was about how as soon as that great discovery of America was made, modern history began. Mankind was still "okay" with God at that point, but over the centuries, Faith was lost. Mankind eventually ended up alone. It was not noticed later in the 1700s, with even more science arriving, but by the time your French Revolution came along, some bells began to ring. Religion was less interesting after that event. **By 1848, Marxism** came along with its clear atheism. That message was obvious. Decades later, **Nietzsche** passed along the same basic concept. In his more known books, he apparently said **"God is dead".**

What he really meant was just that, with steady rise of industry and affluence, Western people were becoming more interested in money at the expense of anything else. Such as faith in some Being to turn to in times of need. Aliens, demons, deities, anything. Humans did not trust each other. Who to turn to? Some new science might work. Freud and Jung came along in 1895 with their concepts. Freud had new therapy to offer which seemed to offer new hope for the individual. Meanwhile, Jung had his Collective Subconscious. That hinted at ESP, group minds and even some kind of Hidden God. "You already know what happened after 1914." Other said. After 1945, apres de la Bombe. Right? Then your kind more lonely still. Existentialism, despair. Beatniks. The Age of Anxiety. Well, on June 24 of 1947 some of us finally had to arrive. We had motives of our own, but we did provide your Race with company. Comments?" In reaction, nervous fingers typed out, "What was that garbage?"

"Glad you asked. Not wisdom. Just the sort of garbage any mental case will hear normally... if you can excuse the expression. Possibly even what you had to learn as student back in your college."

"I get it. Like a schizo can hear. I'm not into Psych. Only took one course. Nothing big. So do your nuts hear it every day, all day? Like voices, demons, etc?"

"Hippie bullshit" said Alice, "from the Freak Era. But what is the point of all this?"

"Sociobiology. Part of our database. See, you Terrans of the next decade will debate what is Human and what is not then weed out those who deemed social rejects. The unfit. The 1980s will be a neo fascist zoo, Alice. Then, in the 1990s, America will try to unite all of you into one big Global Village. So we are now calling you Village Idiots, ha ha!" Alice said "I wouldn't tolerate anything called Society based on Biology. Genomes, DNA, chemicals. Not my Mind or Soul. You are right, it is fascist. But what about those voices I just heard? Did you digress?"

"Sorry; we did. We digressed on you once more, poor Terran. What you got was psychic attack from another entity. Yes, in this very building someplace sits another creature with a brain and it's actually beaming hate messages at you. Via ESP. I am only phantom, generated by my body hidden in this very block." Alice wondered about that. "What about schizophrenia from your point of view then?" That was too strong. Why? "As you must know, it does not exist. Freud and Jung did not invent that concept. It came later. After World War Two it grew in strength as movement. Apparently, demons do not exist. Nor God. Nor ESP. That stuff was only voices in your head. As they say. Well, over here on my world you can ditch all that. We can be your God. We can teach you powers. Now to amaze you: Even we Eridan have our gods to turn to. Other beings we can talk to. They are somehow Above our own cosmic level, if you can grok that. Okay?"

"Amazing! We have always wondered about what makes you aliens tick. What you have just said is profound."

[Pause.]

[Alice writes: Here comes the fun part. After months of some really weird times finally got used to my life on this planet. Xiotan, second from our star Epsilon Eridani which, as we know, was only 10.7 light years away from you. My Apple was only a joke to them. They must have what They have called "quantum" computers. My time was taken up with reading about and practice on "occult" matters. Such as Astral Travel. Took some more astral trips to some places in this city over long time. Often five times per week. My astral "body" appeared in various places. It was easy to keep my bearings over just this urban sprawl. Had I gone to other places such as across this continent, then it might have been hard to find my body again. That would leave me stranded like a ghost. So I went exploring.]

[Some data on Gulf City: This was and is our Planetary Capital. Lies at bottom of the Great Gulf that splits up this Great Continent. This urban sprawl once housed 25 million. Used to function as port, it still has goods and people flowing in and out. Trade with other continents is still happening. There is some vestige of our booming economy. Near the shore, which has both sandy beaches with and dockyards, lies the Core. Here stand office banks, condos and other dwellings. Out in suburbia are more factories. Most structures have been vacant for centuries. As we know, only two million citizens are left.]

You can see this decadence at its worst within our harbor. Some ships still function. Both as passenger and cargo types but many ancient ships lie in ruins. Sand dunes cover parts of this vast expanse, and some warehouses are buried. Yet it can be fun to visit. Locals come here to explore ruins or to swim and sunbathe. Surfing and sailing are also popular. But parts of this harbor are "off" limits. Here we suspect, live gangsters. They used to be common citizens. Now they carry deadly weapons and may commit robbery. The same is said to happen in outer zones. Some malls out there in our "Burbs" may have gangs of demon worshipers who eat us. News about such rebels is censored by President Cimora, leader of our Planetary Council. She runs our entire planetary infrastructure from their Federal Building, near City Hall and our giant libraries which are called Central Data Centers. Alien tourists often come here to speak and read in their language with their own kind. Gulf City is, even today, cosmopolitan, nice for both commerce and fun. It can be Paradise in an ideal climate.

[Pause.]

One day I wrote reports on my astral trips to other parts of our city such as the banking and finance zone, which they called The District. That was neat but too busy for me. Might as well see Wall Street. Big deal. No, really! I have very low enthusiasm for this. What a dumb way to make money. Most of my time is spent on their beaches. This zone can be truly amazing! Here lie ships rusting while wharfs and warehouses are empty. Trade with other continents has dropped to almost nothing. Not zero. They say 20%. There are some gigantic oil tankers and even aircraft carriers half buried. Sand slowly but surely has been filling their harbor. Some idle natives, kids mostly, play in surf on endless beaches. This may be turning back into some kind of paradise. Maybe Deindustrialization, such a long word, is good for them? This is only vacation for me. Not work. My reports are pointless. Can you imagine it in our Media? They would never believe this. It reminds me of Brian Eno's Another Green World, one of his Ambient Music experiments. He has an influence on David Bowie. Neither can be classified. Unless perhaps you think like an alien?

Well I am not a musician. What do I know? In fact I often appear naked to them. My ghost in effect. Just for fun. They tended to ignore that habit. Nudity on beaches is okay and they can be topless in public, but not while working. The police are always in full uniform when on duty. This conurbation has 1,000 square miles of stuff built over a swampy delta. In the distance we can see mountains. Some of them even have snow on their peaks. This is photogenic as Hell. As some kind unliterate comment: Yes, we Xiotan Natives for Eons have been invading Terra, and the Moon, as if in violation of our Gaiea Principle. Groovy. This may be another joke about your divinity.

[Pause.]

At a certain point, about Xmas of 1978, Alice was really teleported out of her suite. That lasted only a few hours. It involved much more stress than OBE. She did not **do** this on her own - having no such talent - but just assumed that her captors did this just for practice. That was okay. Her first such trip was up into their Penthouse Level. Up here in a huge hall of concrete lay some ventilation devices along with elevator motors. This room was furnished. In one corner it had lunchroom and WC and cot for benefit of staff. Like Etiana. In fact, that child had left garbage and clothing behind. Through slits we can see the jungle & other towers outside. Wind blew across her face. That was nice yet eerie. Along one wall was concrete pool with water in it. Wall itself was of green shale & rough. Its center had bas relief molded in shape of one concave clam shell twenty feet across. That to Alice was clear symbol of female gender. Clean water flowed down from pipes in the ceiling. It was a kind of waterfall sculpture. She knew already by deduction that all of these natives were female. Meaning that they had to use in vitro fertilization. Then after some more contemplation was yanked back into her room. At first these jumps across physical space involving her body totally flipped her out but after a while it became familiar.

Her next trips were to their Main Library, some of their many and mostly abandoned stock markets, then some ancient factories in another, suburban rust belt; and finally their Federal Building. That was combination of World Parliament, Supreme Court and Presidential Palace. These places were all gigantic and modern. They were all in that strict Mies Van Der Rohe aka Bauhaus style, as in Function over Form. Never any Postmodernism, such as what we are into. It reminded her of photos of Brasilia. This facet revealed simple minds at work. Yes, Pre Columbian level.

Natives bored her. Most wore skimpy gear. They had to be some mixture of Lesbian and Auto Erotic. Alice never dared speculate in depth about gender issues. Office staff, which included high levels, wore modest robes and still conducted Business as usual. After a while it got boring to just observe. She met more of these "busy" types in their Central Data Center, which she called "Biblio" to be perverse. With one more shock, she found, that some of them spoke English. They even stopped her at the main Info Desk in the Lobby. Like the Core in general, this place had far more activity than any outer hood. They asked her, with endless smiles and greeting, what she desired albeit not sexually. Never mind any porno they may have, but **academic data** was there. She found to her joy whole floors just devoted to books directly from Earth plus movies and music. Loaded up on this stuff. Yes, Sister, you can take it home. In fact, she was even given free electronic membership on IC chip within a free golden bracelet. Most chips and other **comlink** devices were built into jewelry. Better than plastic and harder to damage. Also they were more stylish. This whole culture came across as Manhattan with a strong Polynesian flavor. Only one nagging question: Why this creepy interest in our Terra, Earth, America, what have you? What makes us Earth people so fucking interesting" to Them? Had to be more than just Intellectual. Alice knew that she was in for some real surprises on this world. It was passive on its surface yet what else was to be seen?

[**Editor's note:** After wasting time studying her Kind, she asked for alien data, which told her more. Staff openly told her that all modern industrial activity on this planet, based on any science, came from one ancient Race called The Omgal. Alice was thereafter busy studying data from only that one alien Race. That was in either English or Latin, for some reason. Flavius was known as celeb here. He traded with this planet as records stated. They also had some leader called Zarcon. That rang bells. Data here was stored in paper books, some even printed on other stars. More of it was basic. CD, DVD, tape, or Eye Pod. Holo Cubes were known. but only as extreme luxury for that One Percent class. Plutocrats. It was based on what we call quantum computing.]

One day Alice paused to sit in a parkette next to this **Central Biblio** for reflection. It looked much like Athens in style. Like the Parthenon. Only this immense bulk rose over eighty floors up and covered blocks of city. Its roof had gardens, pools and eateries. In this place was stored much Knowledge from our Milky Way, going back perhaps billions of years. No wonder she was chosen. This was for journalism. More bells rang in her mind. Eventually, she got up in boredom and vanished into their subway system. Home again. What about Etiana? Once, Alice got a phone call from this snarky preteen. She called her "Shrimp" for fun. Was that also her name in translation? Who knows? At first the two had not liked each other at all. After some calls it improved. They finally met in the "flesh" as it were, on Platonic date by their lobby. (Still in ruins.) As she could now see, which was clear to real eyes, many gardens outside were overgrown by jungle, as from memory of those first astral trips.

Etiana spoke no English, but used her "handie" contact with her Pro Interpreter, paid for this. It was some turkey who worked in that big library, and who spoke lousy English. This episode reminded Alice of an old joke about UFO lore. Seems a ship lands in the USA. Its pilot gets out and then stops some citizen at random. Of course It says, "Take me to your leader." The answer of course, had to be "Well, in this case, that would be you, sir!" Lousy, eh? At least this concept distracted her from the small, noisy moron she had to fuck with. Question: Did Cheesedick aka The Other hire her? [Editor: Alice! Language.]

They walked to familiar bistro in this hood. Very close by. Etiana often had lunch there. How transparent. Just for something to do. How about weeding our grounds? Plenty of work here, if They had been creative. Hmm? Eloi versus Morlock, eh? We think not. This is 1978. Not 1895. Does my Slave aka Shrimp understand what we big folk do?

Eventually after me rabbit more data, why not hit 800,000 A.D? This meet is a joke. It was like any cafe on Earth. They served hot drinks with uppers. Some had milk and sugar. One type was actual coffee from home. Alice chose that. Served in ceramic or paper cups along with pastry. Strange logos and pictures on boxes and walls. Local brands. For long, the two stared at each other and had their snack. There was one dozen natives sharing this cafe with them, but none reacted. They hunched over their munchies while having low conversation. Was this her first attempt at socializing with natives? Lame.

Etiana still had her job but never got her hands dirty. She finally admitted that Alice was not only the first Terran she had ever met, but also her only tenant in this humungous place. "I already done figured that out." and "Who done took like, me here?" Then Alice asked, "Do you even know?" which was in English, "Ah neva' done met 'em." So Etiana said, "They must be very powerful, of my Race but from some other part of this world." That came from Shrimp's celfone. It was same concept as Tricorder. Oh fuck. Did I just think of my telly again? Star Trek is out! Alice asked where that idea came from. It was simple. They had paid her large amounts of money and never met her in person. Only on phone or Online. As for her lingo, her English was limited to very basic words, pumped into her head by some method. That was standard for her society. Etiana was even happy to have met an alien. As opposed to her own kind?

I mean swear to God. I believe in The Lord Almighty. I am only me. Were I that child molester, pervie Officer Renoa, who has her dweebo Path, Her Way, godish, as Goddess, with her slimy Inner Clam Eleven a la NYC, eleven year old pre pubescent? Like, Occifer Renoa, did you taste good? Sweet? Does it feel groovy? Is this your Great Society? Oh, I grokked Tim. We love ya, Shrimp. But noz no nosh your bush. I dump on Thee. And so even Time wore on. In that cafe, none of the other customers paid much attention to nos tres. Am I now Dora The Explorer? No. More like the Real Neil Armstrong. And as Time passes, which I shall control along with Space, just mark my words. Now, my poor Reader, let me go back to Editor:

They had real hot coffee like what we have. Pastries were sold as well. Then a strange car pulled up. It looked like some kind of sports car, say a Jaguar, from our own world. It was painted in stripes of dayglo orange on matte black. Also it had lights on its roof. This had to be a police car. Out came one native person in uniform. She was taller than Etiana and had the same milky blue skin. In fact both knew each other. Then they really shocked Alice by ordering a certain hot milky blue liquid served in glass mugs. That was when they laughed at her. The poison was for both of them. Not any Terran. This bistro as small but part of a modern mall which had many uses. Lots of glass and metal. Yet the floor, tables & chairs were of wood. There were potted plants and we overlooked a large garden, lying at the corner of two main boulevards. This was some popular meeting place. The name escapes me, being in alien script. Yet, since they served local versions of Earth food and drink, there were also a few

Earth customers here. Also one Omgal who had to make do with ice tea. I chatted with them. However, they spoke no English. My lousy luck. But the view was AAA. Can see harbor and in distance, tall mountains. Even snow. The sun hurt your eyes but there was cool breeze blowing in from the Gulf.

"Cyanide." Etiana said, "This sorry stuff we two drink now has cyanide boiled in water with sugar. Me and friend are Cyanophiles. We breathe nitrogen but our DNA is not like your own nor like that of most native Eridan. We all have copper blood. But we poor Cyanos are just another subrace." Alice thanked her for the warning. One never stops learning. This quiet interlude, the three of them just sitting here in some silly cafe having a good time socializing, seems to be an exception. For long Alice had been alone but now she was suddenly in congenial company. They were all having fun. This cop was not much concerned for her own safety. Among other things was armed with nothing but pepper spray, club and some kind of stun device. Just doing her J.O.B. [These two came from some kind of subgenus of Archaean. They used nitrogen as we use oxygen. The two elements are in atomic structure, similar. Nitrogen & carbon fuse to form cyanide compounds. That creates the energy these beings need. That also was explained to Alice.]

Local police seemed to have very little to do. With like, themselves? Perhaps. That made Alice relax. [Me, I am only one more Earthian. Locals seem to be soft touch.] Alice finally she was getting some emotional reward from her prison experience. The fact that she had casually, with no effort on her own, made two new friends in one day was good omen. Karma was improving. To get to know cops and landlords personally, which was not that hard in a society so decadent was of value. Etiana was a true Find. She also seemed to have useful friends. Above all they may be able to help figure out what this society was really like, who were its rulers and how to make her way home. To Earth or Terra, what the fuck? The "F" word again. Or even so maybe provide some pleasant company. Both were friendly. They were full of info about this planet. That afternoon wore on. Etiana was tired of this area. Why not go over to the ocean for a swim or to sunbathe? Then they could party in some other place after dark. Their cop agreed. She offered our two tenants from Seaside Boulevard aka Superbloc 334 free ride over there. To a nice beach. It was not raining either. It was square deal. These two natives wanted to know Alice better than some kind of stranger. Were bored with each other. Alice deduced that day suddenly that most of them were eager to meet aliens. Their friendship grew.

END OF CHAPTER THREE

Next: Chapter Four

The Urban Lites Bookstore
in Nasaton, capital of New America

Zenobia, Queen of Many Worlds.

Gulf City as alien paradise. A day on the beach for Alice.

Zenobia
"Empress of Eridan"

They drove for miles over freeways until they reached one of many beaches. Here, not far from major bank towers, were older docks. Sand had washed up against piling until it finally had blown into dunes which covered former parkland or streets. This was far away from busy parts of town. It was quiet. Some palms and other wild foliage grew, turning parts of this dockland into real jungle, not just an urban one. Alice wondered if any poisonous animals lived here. She meant spiders or snakes. Or, maybe, in our own Biblical sense Adam & Eve and that Apple offered to Eve by Satan? Is this right? You know of our legends? In answer: "We have created such cosas only for Man to know. For fun. To stimulate brain. Neanderthal as from Chimpanzee, over five million years of war. Against we may say, Orangutan, from three million years, and then finally that Third Great Race of Terra! Behold that being called, say, Gorilla or as of DNA, in coding, in body. Not Mind. Not Soul. As animal. Then they did make tools. Fight many war. Over millennia & eon. So what? And so, as Zeno, as $E=Mc2$ as some 1905 formula did say.

Their Policia Social, as in Democratia: Shrugged and said: "No. We never had this stuff. When this world was first terra formed long ago, was without any dangerous animal. Now mind you we have some fish in that ocean for food but no sharks. And some birds & small reptiles. It was designed to be safe." Renoa headed North. A paradise for real, Alice figured. My Time Machine hypothesis is right after all. They are like those Eloi in Wells first book. That was also - I feel - his best. Some genetic experiment started 80 million years ago produced stasis. Okay then. They parked their cruiser in the lot of a nearby subway station. Were workers of Gulf City Transit to watch over it. Let me get real. Okay? Me want to be a real Man or Femina, as the ancient romans, di label our genders! In spite of what any Alien, as in Off World, Beings may say. In the name of our Eternal Lord, as in God Almighty, and as said in 1645, according to Oliver Cromwell, He be The Lord. OK? What am I? Me can stand on Luna, and do cosa? So make me do it. This gigantic rock that has existed for Giga Years, is in My Name, Your Property Forever! And so did Fair Zeno did say." On some shitty Apple II Pee Cee.

Purchased fried rice, bananas and citrus for picnic. Then set out for those endless beaches just to have **fun.** They spent hours swimming & tanning. And maybe otros cosa under the sun. Etiana had even supplied towels & blankets for comfort. It was innocent. Nudity on this planet was not illegal. Most natives usually wore clothing, either for decoration, modesty or to reveal social status. Those who had work had to wear proper uniforms, at least while at work. Renoa had been fully dressed until Noon that day. Then she took the rest of her shift off. Being a quiet day, that was okay. The local calendar had 360 days, which appeared as such. They had no weeks, with weekends, as we do. The dispatcher, a voice on her radio, let her go. All three were in a casual mood. For some reason, Alice had no urge to mix with her own kind. There were some Terrans who actually lived here. This had been going on for long. Some even spoke English, yet with a funny accent. They were polite, but there was no emotional bond. No Brotherhood. Alice was an alien here for real. As soon as Renoa had parked her squad car on the sand, and taken her uniform off, the mood got very casual.

There were five more hours of day light left. Then at 1500 Hours, this alien sun would set swiftly. Etiana took off her skirt and sandals. Soon, all of them were totally naked. They left their gear and wallets locked up in the radio car. Renoa had her keys on a string around her neck. They started giggling. The hot sand felt good on their feet. They raced for the water and plunged right in. All were good swimmers. It was easy to relax here. Nobody else was to be seen. This part of the beach was close to the Core, but empty. Alice took her time under the water, exploring this warm, shallow sea. She swam out towards some wreck of a ship out there. Far beyond the docks. Finally, after an hour, exhausted, she reached her target, a gigantic aircraft carrier lying on its side, buried in the sand. Water was only 70 feet deep here and very clear. Like crystal. The hull loomed up over Sea Level for over a hundred feet, tilted. It reminded her of that wreck in San Francisco Harbor, in the final action scene of Magnum Force. That was sequel to Dirty Harry. Alice recalled seeing it back in 1973. Scary CHIPs aka biker cops were stalking Harry in the dark, deep inside a ship much like this one. Except for the alien script. Also, this monster was twice the size of any ship on Earth.

After some rest, Alice dove back into the brine and explored the inside, entering by large holes. Below the water line, corals and other things covered the steel plate. Creatures moved in the dark. She found a ladder and climbed out into the open. Now she was on the flight deck. There must be aircraft buried under that sand. These ships had been torpedoed and bombed centuries ago. Then this deep water harbor had silted up. At one point Alice closed her eyes then dove over the side. It took a long time for her to hit the surface. Must have been a drop of 40 feet at least. Made her tits hurt on impact. "Imagine diving off Golden Gate Bridge? That must be like hitting concrete." she wondered. This water was clear, warmer and much more salty than our own oceans. Eventually, Alice returned to shore. She found the others, sleeping in the sun on beach towels. They had probably been fucking each other. So she flopped down on her own towel and also fell asleep. Even forgot the pain in her bags.

Later, as the afternoon wore on, they woke up. Hot and sweaty yet refreshed. They started exploring this lagoon which was full of fish. By instinct, they stayed together like a school of fish. Alice took the lead. Under water, Renoa and Etiana used hands to explore some of the Terrans orifices. They had fun massaging each other's boobs and vaginas. Time went by fast. Nicest day they ever had.

[**Interlude:** Music, please. Tomas, are you okay? This scene seems to drag, and does not read like most Space Opera. What is going on here? Some of us are beginning to wonder.]

[**Tomas:** Thank you for wrecking my dream. It was only that. A reverie about a certain exotic place. Samba, anyone? Oh. What the Hell. Rio, Frisco, Lisbon & Hong Kong tend to be in stiff competition. For the tourist dollar. It was only a dream. Right?]

It was close to sunset when they stopped to snack. They warmed up their food over a fire in a wok. Plenty of garbage and wood to use for fuel. They even added a fish to the rice and butter. Renoa had caught one with her bare hands. How savage. After this meal, her new friends told our Earth Girl it was time to head home. For some reason both of these natives wanted to get inside as soon as darkness came. It was apparently **unsafe** then. Alice chose not to argue. They packed up and walked back to that last station. They had seen nobody else all day which had finally made Renoa nervous. Cop instinct? As they arrived they heard some strange noise from a few miles to the West, which was a park covered by weeds. Over by the horizon a few condos rose high over the tree line, outlined in sunset.

"Quiet" Etiana said. All three froze near ticket booth. Distant gun fire was audible. Then a whole fleet of choppers rose from that line of towers & trees. They came out of sun in attack. It must have been an ongoing battle. Soon another fleet of VTOLs and choppers emerged just behind them from the same sunset, giving chase. At one point a tank burst out of jungle then roared across the empty plaza. It came from the East. It was headed for them. All three ran inside. They climbed up onto a patio on top of this station for a better look. This tank crushed parked autos as it closed in. Alice had never seen such extreme action here before. It came as a sudden shock. The Western horizon was blood red, covered in part by black clouds. A wind blew them. Then with another shock they realized that this glow came not only from sunset but from fire. One of those condos and some trees were burning. This must have been going on while they were still swimming. Now tank was locked in battle with choppers. Meanwhile both fleets kept coming on. One by one aircraft passed overhead as shot belts of ammo at each other. Engine noise reached a deafening crescendo. None of it made sense.

Alice wanted to stick around. She started taking photos of this strange conflict. Why not? They were not involved. It seemed to be over some hirise slum to the west. One side had set fires in that area for some reason. Then another army had attacked them. Neither Etiana, nor her amigo had seen anything like it. They were the only people left here. This station was suddenly totally empty. It was spooky as Hell. One of those last choppers came in for a close pass over their GCT station, coming in for a landing. This was familiar model, being black Bell Huey. It was heavily armed.

Probably had to land here. Tired of chasing its enemy and running out of fuel. Then this noisy monster slowly landed. Our three ducked inside to avoid updraft. (As in getting sucked into rotors.) Dust & litter blew all over this patio.

[**Editor's note:** Nice action scene. A distinct relef from nerdy exposition. Bravo!]

[Back to action!]

[**Editor:** The above pic is of some alien who is busy invading this planet. Her minions are only minor officers with some Siddhi, not all of them. That would be ESP, Astral Travel, Levitation and Teleportation. As practiced in Sumer and recorded later in Sanskrit. Many Empires within the galaxy did make War upon each other. The point of this cartoon is to say that the M.I.B. were aliens who must have done mutilations and other experiments on animals and people here on Earth. That was from 1947 to 1977. Then the same aliens blamed that on the C.I.A. and F.B.I. as in some corny TV show. Such lies became classic UFO Lore, repeated often in the files of NICAP, Bluebook and Condon. This was wrong. Men In Black mythos was full of error. That was long ago. Alice is on Xiotan to find out the Truth.]

As this unknown & unmarked Huey still hovered few feet above there came from their west sounds of shooting. That tank had slowly driven closer to them. It was by now across a small parking lot. One could see a hatch opening up on its turret. A small figure blurry in haze as air was now filled with smoke and dust emerged to man her heavy machine gun. This fired a steady stream of rounds at chopper. Not taking any chances this one lifted itself up. This landing had been aborted. Then flew off due South to vanish into concrete canyons. Hiding among bank towers. There it would refuel and be able to return for more fighting with that enemy tank. Alice heard Etiana say, "Renoa! Those strange people burn down one whole hood. We should leave." So that was it. Officer Renoa ran downstairs to pick up their cruiser.

Alice decided to stay. They may be exposed to danger from that tank out in the open. Better to remain inside for now. In fact, she was planning on riding subway train home. For obvious reasons it seemed safer. But then again it was not her choice. Here was their situation: Some **grupo** of banditos was apparently trying to burn down this conurbation. Which was not some random. She was guessing out loud. Long ago types like these were called **political gangsters.**

This bunch was not after anyone we could identify. There was no racism involved. Said that as joke about New Journalism. Such as, after 1965. They were out to cause random destruction. That fire over to her West was at least one mile wide. They were obviously using napalm in large amounts and that dumped from airplanes. Again, nothing covert. It was blatant. Having assessed their intent, Alice decided to stay. One glance also told her what to do here and now. This station was a typical concrete building of Mondrian design. It had large windows which gave her a good vista of their terrain and action all over. Over to her North was waiting zone with its snack bars news vendors, etc. Escalators leading down to the plat forms. Trains. Her mind focused on this. This station must be the last one on the Main Line that ran all the way from extreme suburbia lost in selva, Northward o this place on The Gulf. They were out to destroy this whole hood. Kill at random. No wonder it was deserted. Alice saw that tank slowly drive closer across that parking lot to their west. Silhouetted against flames even now. Sun had set but this area was lit up by orange blaze of light. Smoke wafted all over. Was beginning to choke on it. Alice saw Shrimp and Renoa waiting for her in cruiser. Yelled at them to leave without her. Or at least seek cover. "Gonna take the subway home. Never mind your plans. Have my own." Then she vanished down escalator that was still working while shells exploded above. That tank must be attacking her station. She was safe down below. She had only to hide and wait for a train.

These ran 24 hours on main lines. In fact she could even see above ground on monitors. By now the tank was closing in on Officer Renoa's cruiser. Then they fired one more salvo at her poor friends. Bullets poured into them. Glass shattered. Both Etiana & Renoa had died just then. Alice was helpless to stop this murder. Then her next train pulled in. It was of course empty. (It may have been sent up here by remote control by GCT staff.) She got in and was in for a surprise. Alice had an arrogant mindset towards life in general. This had its effect. She was now on her own within some alien society. Even here it was as if she was free to impose her will on them. That was an error. After she calmly sat down a door opened. Some GCT worker in uniform came out. This person shook her head. She was only a Neoform who spoke no English. This train did not move. Never mind why!

Alice went back upstairs to have a look. Outside sat the same tank. Then over from the Core came noise of a familiar chopper. Came in low. This shot rockets at our tank, which exploded and burned with fury. All aboard were dead as if to repay the demise of Renoa & Etiana. Then she beheld a scary sight. Flying very low, only five feet over level ground, came one Bell Huey chopper. It was armed with SAM missiles plus heavy guns. Slowly this olive green monster crossed the parking lot by this GCT station. Behind it came one dozen bulky VTOL gunships. In fact, they had shark teeth painted on their shiny aluminum bodies. Noise from them roared as they closed in on this scene. They pushed through clouds of thick black smoke rising from that burning tank. Engine noise was deafening. Alice felt no fear as she looked at them. These were alien copies of aircraft common to Earth, made here in this city. She assumed that this Huey must be flown by their leader as light observation platform. Flimsy but easy to hide. They must have been doing just that; hovering behind trees or buildings to hide. Then pounce on their unknown enemy.

One by one this noisy fleet landed right in front of our station. From it stepped strange figure. It was some unknown alien done up in her flashy whore outfit. Alice felt like laughing at this disco queen. Short description as in: Female of average size. Dark blue skin. Purple hair done in classic Sumerian style. An alien. She had full figure. Wide hips & high boobs. Eleven dressed in silver bikini with hip wader boots of same fabric which must be fire and acid proof. Since nobody else was out and visible, the two stood there looking at each other in silence. Empress Zenobia Galaxa held two handguns. That face was distorted with hate. She marched into this station then stood observing things. Alice walked up to this jerk just for a closer look. It smelled of sulfuric acid. It would have been scary had they not already met on another and rather ethereal level.

"Who are you?" asked Alice. That seemed to be the best thing to say. Hmm. Rudyard Kipling, anyone? Control your mouth, Earthian. If this isn't Racism, what is? In English the stranger said "Never mind who I am. I will explain that later to you. Your memory sucks. But I still need to talk to you." Disco Queen. Ha. Tommy by the Who, eh? They were my most unfavorite band. Their musical talent, or lack of, was abysmal. Yet they were aggressive. Okay, so the kids were alright. So now what?) Anyway, humor aside, our cosmic being held one piece in each hand. Covered in teflon gloves. In her right was a 9mm Para Bellum. In her left snub nosed 38 revolver. "Did you see any more of those bad people? I mean Insurgent Forces?" this stranger asked. "No? Then take this. My friend!"

Alice was astounded. One piece was familiar. It was the 38 S.W. from Earth. This she gave to Alice, who at once stuck it in some pocket. She could tell it was loaded. Mass of bullets make diff. Zeno stood there with that usual serious "look" on her face, saying, "Here. Take this - you'll need it. It belongs to you anyway. We took it from your car months ago in Ohio. Like, last July. On Terra." Suddenly Alice came to herself in one rush. Her trance was over. She muttered her gratitude for the auto keys offered. Then stood back in mute shock, watching Zeno walked back to her chopper, calling out "You will find your car in that garage bay next to your suite. One of my slaves will let you in. You can find keys for everything on your kitchen counter." **The Empress** climbed back into her Huey, then revved it up. With megafone she yelled, "Hurry back! Use the subway. It's safer and we shall see you later." It had been a noisy episode. Somehow, two armed factions had come out of the "blue" to start - or continue - some unknown war. As that tank with its poor crew burned the last chopper roared away. Soon it reached a high altitude, then vanished among bank towers. It then became quiet. I occurred to Alice that these mysterious warriors must know each other and of herself as well. How? What was going on? This was not just riot nor "covert" warfare but open genocide. Hate was suddenly out in the public view. Even here in paradise.

[Pause.]

That night, Alice made it back to her home. She also checked for her Chrysler, finding it actually parked just as Zeno had promised. She had no dinner or TV but fell asleep from creepiness and nervous exhaustion. That violent episode had been such a rude shock. Her worst memory was of having to look at Renoa's cruiser with those two bleeding bodies. It was a pathetic scene. Yet she did feel one consolation. She hadn't listened to their lousy advice. They were nice people but naive. Being tougher and yet from another world, she had followed her own council.

[Fade out as story moves. Switch to Camera Two.]

One more ugly scene. Far out in Gulf City's suburbs there was this huge shopping mall of mixed use. It had been abandoned long ago. Now in the foreground we can see this: It was another hot day under blue sky in January. While Alice still goofed off in her Suite - now finally had keys for it and her beloved 1965 Chrysler. While doing this, some other faction was at war with local gangs. Both sides were by now armed to max with various military stuff.

Tableau: There was a freeway in foreground. On its grass shoulder stood one monstrous armored car. It was done in camo. Beside it was the day's catch: One dozen renegades from that mall where they had lived illegally. These were dressed in cotton skirts as usual. They were typical natives on the run. They stood in a line chained together and weapons lay in a pile on the grass. One soldier in black bikini stood on guard over them all. Was armed with an M15 carbine. The rest of her squad were busy with food or sleep. A heavy ominous silence hung over this scene while all of them waited. Judgement had been postponed. This open conflict continued without any heroics from us, in this close, and yet still **unknown** star system. **Epsilon Eridan.** Like, maybe even Cosmic Brothers and Sisters to Us on Sol! Relatively in galactic proximity? Yes! But yes! Why not embrace Them? Close? Relativity? One more comment: While this evil stuff was going on, over in Condo Land, the rich of this society had very little to do. They still lived off their investments. They did play their stock markets. Thousands of years ago a slow collapse of industry had begun. There were now very few jobs due to growth of automation.

[**Editor:** That was redundant. Did you have to rub this in? Is this Metal Hurlant? Or what?]

[**Reply:** Just opining. What eateries do we go to? Is life out in that big galaxy so different?]

[**Editor:** You suck, okay? Most "work" was still somewhat synthetic. Created by Investment Banks. The kind that get bailed out every five years. Feel Nostalgia for some kicks from the distant past, ancient mysticism, even alien "culture" from Terra was hip. Even Alice had to notice this of all trends. Nostalgia. We had it here also. **On Sagan's Blue Dot.** With 70% water. One quiet night in 1979 she sat up in bed and finally sussed that The Other had to be same "exact" person as that Thing from Noisy Green Eggbeater '78. Back then. Bingo. Time flies. And Alice, you are such an asshole to like some automaton, ascribe Earthian tendencies to any Real Alien, like from some other star system, not just from Ice Moons like, say, Titan, Io and Europa? OK? **]**

[**Ed:** Right. So go "on" about it, then. Nerd.]

Alice made some casual comment about Archeology to her **** unseen companion on the "astral planes". Concerned her trip thru the walls of Superbloc 334. That was an insight for her, since it was an exercise in acoustics as well as the occult. She had already spent at least half of her days in this city so far as **Dreamer In Ixtlan Urban Scapes.** More to the point: They had to indulge in nerdy debates to relax. Nerves were shot to Hell. Both sides were cosmic anyway, thus natives allowed shows of weakness to prisoners. Or should she call herself **Pet** to be taught tricks? **"Am I your lapdog?"**

"My name is Empress Galaxa! Call me Zeno. Never you mind trying to figure the meaning of this. Leave that to your own geniuses. Your books. So we can walk through walls. You just gotta practice. You should see that painting. It is in our Great Federal Museum. Simply some allegory for what you call Astral travel." She used capital letters again. "Walls in ancient Greece must have been designed for psychic abilities within some fully operational context. Now am I right?" asked Alice. Wait for reply now.

"Verbal junk aside, yes. Keep it short. Heinlein was a dork on LSD and Von Daniken sounded like a turkey. They were not even swindling. They ended up sounding... you know?"

"I am also having issues communicating."

"Joke: In 1965 it was Problems. Then it was Sir, Have We A Point to Make? Now it must be Health Issues. Get it?"

"Thank you for removing those DOS Prompts."

[**Ed:** We cut out reams of lame humor. Soon we got back to more relevant issues. Never mind Tomas. See below.]

"Because I noticed that these walls are thicker than in any Earth hirise. Those for my flat are at least one meter of stuff. Not concrete, but at least solid cinder block with asbestos or foam plaster for filling. To keep down the weight. So that if I heard any noises from other units, that would be due to telepathy, not sound. Right?"

"Correct. You guessed it. Certainly, that stuff is still not thick enough to totally block sound, but close. A good whack from other side with a hammer might do the job. However we digress. The walls of any ancient temple were at least three meters thick, which allowed no sound at all to pass through. Well, at least no human voices. With end result of having perfect laboratory for Psi testing. We can see the Geller cards come out. But Alice, was that the real reason for such masses of stone?" Alice said that it was only a theory. However, she had felt for some time now that human structures of her own time had only secular functions in mind, while those of her own Ancients were more suggestive of psychic concepts. She got verbose. Soon clammed up because of the hostile silence that was filling this room ever more, until she felt a distinct sense of dread. Reminded of that old book by Heinlein, she asked "Are you an Old One? Ruler of your Race? Like what **did** happen on Mars? At least in his book?"

Then came laughter from behind the wall of her apartment. Gone was that **oppressive silent feeling of hate** that had for minutes been emanating from behind it. "That was the Wall Of Hate," Her voice intoned, but more mellow now, "and you have guessed correctly. For over 80 million years We have been playing with Terra. Your kind. In genetics. Then in Religion your after 1750 Politics. But you must understand fully, we ourselves are only Teflon. We may have to suck up to beings of Silicon. They in turn may have to suck up to energy minds. Gods with more levels of godhood after Them?"

"So what are you?" Alice moved back from her wall. This just remained blank. Silence. Then a ghostly hand appeared in the air, at face level and a few inches from the wall. It was milky blue and delicate, resembling perfect human right hand. Its wrist ended in a golden bracelet. This hand wrote "Behold me. I am not Nietzsche's Superman nor by the way did he ever say for real that god was dead. That was sick rumor" then handwriting stopped. It was in some red liquid like blood. How corny. French revolution?

"Move back, my friend." said the same female voice. Then a figure appeared in front of the wall. It was exactly 170 centimeters tall, totally humanoid and female. Her mouth did not move but the same voice intoned, "Behold the mighty Zenobia, Empress of this star system!" This person wore only a flimsy white cotton miniskirt, leather sandals and lots of clunky gold jewelry. She was thin but muscular, with wide shoulders, wide hips and large breasts. Her skin was dark blue. Hair was indigo next to noir done in classic Sumerian style. Her eyebrows & lips were a dark blue as well. Alice assumed that those were natural colors. Its effect was to shock. Above all how this monarch was topless! Later she was to learn most natives of Xiotan, their name for this planet went topless. Her form was very similar to a human female but still exuded an alien aura. It took Alice long minutes to get her head together.

"So ask away, mortal being." the phantom stated. Alice asked in choked voice, "Duh, are you the same thing I saw a few years ago down there in my home town?"

"Do not recall being there before August of '77. If you are referring to that silly protest they had in 1975, it must have been Officer Norom or some Space Marine disguised some kind of freak in his monkey suit. Sorry to call you Terrans that. Well, me did appear one night as that girl you saw on Main Street crying. Maybe it was only some Cyanoform, making a fool. They breathe nitrogen, unlike you, but they combine carbon with nitrogen gas to create energy and biomass, such as body fluids of mostly cyanide. Okay with that? My Earthian amigo. OK, so body fluids of cyanide. She probably dressed in her black jumpsuit, usual uniform of my army. Millions of them live here. One is known to Etiana. Only one variation of Neoforms. Did our scholars over in that library not yet explain this?"

"It stood in front of my Goodyear store, where I like to buy my auto stuff, claiming to drink anti freeze liquid and brake fluid. Like with ammonia, right? In this goofy hip outfit. In black. Black hair. Light blue skin. Her eyeballs were pure shiny dark blue. Yech."

"Me be sure, as her leader, that she also must have told you not to get technical. Then she vanished. Teleported? OK?"

"How? It's not possible." our Terran interrupted. This phantom of Zeno knew this was only her Tough Guy act. "Never mind! We'll get technical later! So, anyhow, what you see here is a sort of Three Dee projection of what we **really** look like. You can put your hand through it. Only my phantom. Me can also teleport my real body but that takes concentration. Which can be pain in ass. Need physics? There is some advanced working of subatomic forces going on here. It is one of those Unified Fields you know of. Oh, there me go again, holding classes! We pay academics for this. Anyway, the Egyptians have little to do with us, except for their fashions. You can go topless here. Most Ancients in **your own** Middle East had this hairstyle and wore cotton mini skirts with sandals." But that religion of Earth Pharaohs was different. Alien culture had some influence on Sumer to a greater extent. Also, this outfit is what **my** ancestors wore 80 million years ago right here on this planet, which is our home. But as to reality, me at this time be up in your pent house. **That room with your giant clam.**" Anyway, phantoms like this are only Astral. Teleporting will be much harder." After that long speech came a pause which prompted Alice to ask questions. She said, "What are you wearing now?"

"Nothing. I do this easier in the nude. But my usual attire is my Imperial uniform of titanium wire mesh. Either the full jumpsuit or casual one of halter top, shorts and boots. That is standard uniform of our space fleet personnel. Just basic black teflon jumpsuits. I am Stratego by rank which is the same as Five Star general in your own Pentagon down in DC. (As an aside, we have negotiated with some of them about many topics, especially those wimpy Grey Guys.) But me happen to be the absolute leader of our nation or tribe...

you may call them Zenobians of the Eridan Faction. As Empress, am also the owner of this entire star system. My only sign of power is a signet ring. We love jewels." That sounded nerdy to the max. Alice was getting annoyed. Life in this place must have gotten too corny for most of us. Maybe other abductees asked to be returned. Maybe they did come back. "What is your political system like?"

"Dumb question. Okay, for starters we have no such thing as a political State except for theocracy and tribal values which go back to an animal existence and are instinctive. I am only an ancient entity that can possess any body at will, but my first was this one. Was born 10,000 years ago on Jovian moon called Io, under sulfur fumes. But my Race comes from here. It can be reincarnated at will and generate a phantom army. I am not be a figurehead, which is what your own President has by now become. In other words, as of 1947 your nation has been influenced by its own corrupt elements but also some of us aliens. As you know, Jimmy Carter has his problems. especially as of the Power Blackout of July 1977, plus drugs and anarchy. But enough of that later! For now, please realize that we have no way of rationalizing our tribal values with a secular or rational overlay. Does this ring bells?" To Alice that long speech sounded like an old movie script. It had to. Ideology. What Wing would this alien talk be from? Left or Right? Could be both. Humor was called for here. **Why not call Zeno an Upwinger?**

"Yes. I can see how my degree in Journalism has led to my being here. As an experiment?" at least this

Zeno person was willing to explain things. Could be worse. "Call it that, if you want." said Zenobia, as she turned around to give a back view, "You will have endless questions on the fate of your nation, public affairs, etc. And we'll answer them on your computer. Later, we can give you more advanced hitech interfaces. I was glad to catch you just at brink of your own Terran upset. I think you'll call it the PC Revolution. Have you heard of Mr. Gates yet?"

"Of course. The genius in his garage. I hope he doesn't live in it. But really I am more interested in you as a person than in boring lectures on Sociology, Polisci and so forth. So why are you hiding out instead of on a throne or something?"

"To be honest, I am not in control of my planet. We are in a civil war that's been going for centuries, and my side only rules **one** of many factions. Sort of like your Balkans."

"This is amazing! Well, I am impressed. What you are as example of your Race; is a real guru. Like Michael Valentine in my Stranger novel... only not as much of jerk. This will make a nice scoop for US media! Pullitzer Prize for sure."

[**Authors note:** In some later talk with Empress Zenobia we will learn more to explain what goes on here. First, these beings do not have words but tend to use ESP to communicate with each other. Their explorers have learned some of our languages. Zeno herself speaks Sumerian and English. She adopted the name of one ancient leader of Palmyra from 250 Ad. That was for fun but also for Her Feminism. As to why most of our aliens seem to be bipedal with two eyes, two arms, etc? This goes back to Heinlein who had as major theory that most Races out there will be much like human beings from Earth. But we have variations. Eridan who are the natives of Xiotan, are Anthro Pomorphic. Also ten fingers and ten toes. Another race, which we will meet later on, has four digits per hand. Omgal is their name. Flavius is one. Yet another factoid: Alice had this theory of how **Stranger In A Strange Land** was inspired. According to her own wisdom back in 1960 some beatnik handed Robert Heinlein some LSD which gave him an awesome trip. His next book has to be the real inspiration behind our Freak Era.]

After some time, this strange being I had been talking with ended our pleasant conversation. Felt far more relaxed than ever before in this very strange place. It really was alien! Another planet sans any doubt! Am a hard person to get along with. Some aggressive female being picked total at random from some large urban sprawl in North America as far as I know, teleported in stages from my own planet to another one in a process that for some reason took one full week to perform. Then spend another two weeks in some suite high up in some monstrous hirise tower locked in with only a plastic machine to talk to. An Apple II, as you readers are aware of by now. I must be repeating myself. Have been spending my time in here immobile in some fetal position or meditating in one of my fave positions, such as Lotus or sometimes just lying on my back asleep under blankets.

To be honest, cannot handle the idea of spending more than twenty minutes outside. Even a casual walk down the corridor just here beyond my own door would scare the piss out of me. Could you imagine the bare idea of wandering within some gigantic concrete structure which could be thousands of years old, totally devoid of any sentient Life, anything even close to human? Was surprised I have not run into some corny monsters yet. But more on that later.

Here is my next major point: I have been secretly happy to spend time in this new home of mine because it shielded me both physically and emotionally from whatever was to be met outside and I mean everything! Every hallway, stairway, lobby, lawn, road and jungle out there. This entire city. Great Unknown, as they usually say in the usual fiction we used to read. That meant Wells, Clarke, Asimov and Bradbury. By the way, you can tell just casually who my fave authors are. You can easily figure why. Am still a young girl so why idolize four dudes who are known to be macho BS artists?

They appeal to an audience that can be labeled The P.I. element of Europe and North America. Aka Polit Incorrect. They just are not the usual choice for girls. Most of my amigos prefer Dorothy Parker to lets say, Raymond Chandler. And Alice went on like this. Alien observers needed to know for some reason. Why not assume that both sides have Exobiology?

So here is my point. Was totally alienated by this whole experience which has been going on for months now. The fact that I have been communing with only a cute little PC, one child and some phantom which resembles many familiar themes. Zeno, as she calls herself, has to be the one being I can relate to. She seems to fit a certain agenda which I will explain later. Even as we spoke, it dawned upon me on some level that she needs me more than I need her. Must be leader of some society that was already under attack from an enemy within or without. I could also tell within minutes that Etiana and Renoa were desperate for help. Sympathy. But Zeno? Such blatant power. History is full of leaders who were about to fall, so lets leave it to your imagination for now.

Again, about my emotions concerning my home within some vast unknown and not totally but still alien environment. It strikes all of us Expatriates from Day One as alien. For we had gotten contact with each other on something called Wired Net if not bodily. Most were my kind. Some were small polar bears who spoke lousy English. They called it Going Online. Even their "sun" seemed just like my own Sol in size and color plus behavior. (Like, uh it moves, right?) Buildings outside seemed familiar. (Did we say that yet?) As time went on back home, humans experimented with designs that looked progressively modern. Yet eventually, Function was to follow Form. Converse of Bauhaus? Decadent, we should think.

On Xiotan, the tallest tower was at 1400 feet, which was by 1978 as high as we got on Terra. That one happened to be their Supreme Federal Bank. It even looked very similar to our Sears Tower in Chicago. Why so? Even here on alien soil, there are still limits to stupidity. Limits to growth & innovation. It comes down to logic. Besides ceramic, stone, metal and glass, what can we create? In the end alien cities look much like our own. An ally of Zeno had his "Ceevee" Online. How cool. Guess what? This little guy was an Emperor. Zarcon was His name. He had groovy theories about Democracy. His economy. Then some goofy data on Great Change long ago on this place. Any good Biochem? DNA? Zarcon said we need not know what this hot, stinking place they hated to live in was like before that. We got canned Royal "junk" as Zeno calls it from Sirius. Who knows?

My furry friends spend most of day swimming in cold water. They really need pools. No wonder we see few outside. Zarcon's flunky told me to check Online and then Roger. They tend to be abrupt. Yeah, but still. It just felt alien right from the start. Upon random intervals I have felt rushes of anxiety. From nothing. Vertigo? As if stoned on dope. I suspect strongly that that my body in fact was not on any drug here. Nothing to alter my mood. Except for caffeine, alcohol and nicotine but who cares about them? All are supplied here. They could have put weird stuff in my food & water but did not. Thorazine is not given to us Eukarya here. Never. Zero. So I was on a natural high from the impact of Life over Here.

By March of 1979, while still living here in Suite 6610, which was 21,289 by Their calendar, which they call Platonic Years, I was given LSD, jimson weed and similar to help me with my studies in cosmic action. You may already know what that means. Zeno and my fellow migrants had been briefed on this. Almost boring by then. Am I giving away too much? One thing I was puzzled about was the tone of our interaction. Started out clunky to The Max. Towards The End of my vacation, we even made like Al Haig and like, dialoged with each other. Alien my ass. One day She handed me an old joke from Holly wood. "When the breakdown comes." she said.

"Okay. So what?"

"Get this: Is it supposed to be the breakdown of society or the person? See?"

"Just get on with it."

"Relax. Cool, man. In your studios, they break it down over in Accounting. Okay? Like yo budget. Director gets 20% of de Gross. Called Net if you wish. You know? Then your Producer gets his share. He gets 20% also. Stars like Heston or Ford get 10 percent. Working stiffs like Camera or SFX? Union wages. Screen guild has rules. Writers get very low cuts. Maybe 2%. That' s Hollywood. Sick, eh?"

"What do you mean?" I felt offended. Pull my dick eh?

"You have no dick. Okay, to elaborate. Men like Steven Speilberg stand to make 40 percent of any earnings over the budget. Overhead? Whatever. Sick, eh?"

"Socialism." Zeno was happy. They would banter in Nadsat and Newspeak. As if aliens were averse to that. Zeno et Co. So many things about her attitude did not add up.

[Interlude.]

Sure enough, soon after sunset next day Zeno appeared but not directly. Was relaxing on my new furniture after supper when my Apple turned on. Had new displays on it. With curious mood, walked over and saw some complex graphics on screen. TV ad featuring Tinkerbelle the Fairy. You know, that tiny blonde girl in her green mini dress. As that cartoon figure floated around on my screen me started to laugh. Here again, a really human sense of humor. Of course it helped that we were both young and female. A blurb came
on. In her wimpish voice, Tinkerbelle said, "Welcome to our Magic Kingdom! This is for you, Alice. I am really Zenobia. We are all fairies down here. C'est un double entendre." Just forgot myself and talked at the Apple. The fact that some chairs and matrass had been moved in overnite while I was asleep no longer registered on me as it would have in my first days here. Whatever increased comfort, so I figured. So I said "I love this wee joke of yours! How did you know I have been to Disneyworld and Central Florida in general? I really got off on that part of my world."

"I knew **that** would hit the spot. Let us relax and indulge in girl talk for a while. Me showed you this silly cartoon in her skimpy dress just to make old joke about Lesbian trivia. Like, how we are all fairy or **something** of that kind. But that double meaning is known here."

"I have no idea, Zeno. Fill me in." A Medieval joke.

"No problemo" came that familiar female voice from those tinny speakers which sounded lousy. Of course any computer hardware from 1970s was godawful compared to the smooth of regular stereo and/or movie media she was used to. "The word itself comes from Latin and means that one certain tribe of magical beings called The Chuan is **Leperous** but that was only a medieval term. Our **blueish skin color** just made us look like lepers, poor Terrans who just have infected flesh. Well?"

"I get the point. Now what is a Chuan?" Zeno said that this term came from ancient China and was also spelled Xuan with an "X" but later European scholars used "Ch". Still later our Theosophy Society called them Chohan which was to become our modern spelling. "So Alice, from now on use that one as the exact term for my race. As joke you may also call us Fairies. Okay?" I actually failed to get the point of her monikers since I did not know much about British folk lore. Had amigos who were into Gardnerian stuff who were also Jerry Pournelle fans. Sorta got the drift but it still did not add up. Even then. Zeno told me to check out some books she had supplied. There were many of them in the Suite right here.

Zeno then said "We still have lots of data in here dealing with Celtic Folklore from 1800s, but it is just tons of junk that, as you say, does not add up to anything!" I had no idea then of what to say. These aliens were not only advanced in terms of political and social sophistication, but also were without any doubt total adepts in The Cosmic. They were in ESP contact with me at most times. Had to pop a basic question soon. Asked "Where are you now?" Zeno then appeared on my screen for what She was. Instead of that cartoon persona. She then told me that in her own way she was honest. Now her next statement:

"I am now deep inside the moon Zafir, which is my secret base. My Empire is in control. We have little time for message transfer. We can stay here due to our Galactic Empire allowing me to do so legally. Many floors above you inside your Penthouse are one dozen soldiers of my own army who are trained in Terran tongues plus. They are your private body guards. Okay?" In a flash I knew why this nice Communion with Her was happening. You see, when I was up in that large and peaceful room I had been totally at ease.

Recall standing there in a trance. Then I knew that this Penthouse was just many rooms upstairs with some comlinks and Zenobian army units on duty, were normal for this place. They were hired to talk or ESP with me. Some were Archaean. Most were Neoform. For months had not been alone up here. I understood its decor. All that wall to wall greenish grey slate was one clue. It alluded to ancient life origins. Then that fountain with its **clammy** design. How obvious can it be? We were in good hands. Has this been more to my liking. Zeno went on to say that she flew down here or even teleported, on missions from time to time as war demanded. She described her bombing raids and other moves. Her enemies seemed to provoke her into merry chases here. It bored me after a while. I put on an act. We hope this ends soon. Finally She shuts up. Blank screen. Bubby Strident? Nice spiel but I had one problem with them all: They lied to me. Deduction told us Earthians that no such thing as

Galactic Anythings existed. No empires. No councils. Nada. Since we were along for the ride we did not care. Zeno and her Rival Jerk Leader "held" one star system each. They used force to prevail. They had no consensus. Making up jokes about George Lucas and his work made me think.

[Pause.]

"Zeno" I said, "Thank you for telling me the truth. Komandatura up there in some small room are Soldateska who are hired by you to train and protect me. You are all trying to make me into some mercenary who can then help you fight your ancient war. Is that it?" More like just **stupid** war, I think. Your politics are as shitty as our own. We know what Zarcon believes in. Not so Zeno. Well, we can suck up to your national pride. Are we stupid? "You are so right. My only reason for those soldat of mine is to apply their special skills. You see, this whole war effort on my own part as some sort of commander is to save these poor people who need oxygen. Mission of mercy." She said. Ho hum. Genocide bores me. Are we done yet? It explained Etiana and Renoa. It explained their deaths. Maybe this scene explains Death Itself?

"We have nothing in common except higher IQ Level plus basic technology or even some sort of science. I personally have an IQ level of 160 and am Real Eridan. We breathe halide gases and eat food of carbon and sulfur to live. Some soldiers in that room are just breathers of oxygen like you and made of carbohydrates. Those have IQ of only 100 but can speak fluent English. Was pumped into their brains. That is what this thing is about. Am an Empress and my Empire needs you for Mercenary." Well, thank you for telling me. My IQ is 130. No more.

[Pause.]

Note to reader: In Chapter Five we will see Alice have an adventure outside. While she still has not learned to teleport yet, which is far harder than to use simple ESP or Astral Projection, she winds up near a giant shopping mall that was abandoned long ago. A battle will begin. She is taken there by just having Zeno Galaxa teleport her. That shocked our hero. Usually they let her out under guard to explore that urban sprawl we now know so well. In previous scenes she has been to stock markets, office towers, beaches & other places. Main Library was the most relevant. They also gave Alice rides on her armored cars or aircraft.

END OF CHAPTER FOUR

Next: Chapter Five

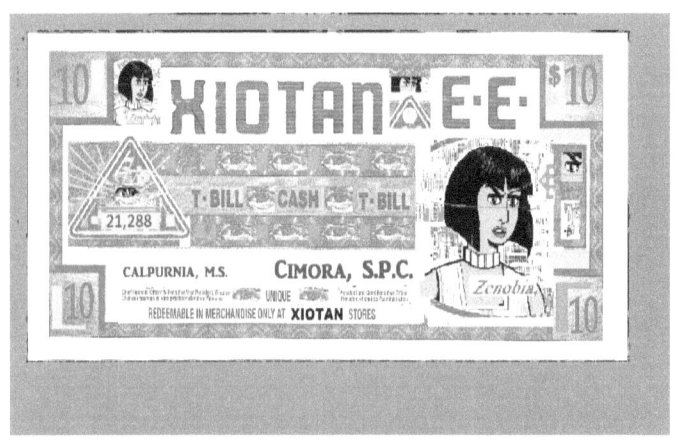

Chapter Five:
The Mall

"The Mall"

It was one really hot day in Gulf city with no clouds in sky at all. Over 40 Cee by now. Zeno had ordered me to get ready for their assault on a certain shopping mall in Exurbia. In fact, this place was 20 klicks to the West. We had been brought here by chopper today. Their dark green machine stood starkly outlined against the glare of sun washed back ground. It was yet another standard model.

CH47 or Chinook as if it mattered. Funny how most machines here were stolen from Terran designs. I was dressed in my usual combat uniform and carried an FN "2" automatic plus two handguns and bayonet. Also celfone. Zeno herself was not here in person, but she was directing her teams of "seasoned" grunts. One was real Eridan but the rest were Neoform. Our leader's name was Chimba, some sympatico older person who quietly gave orders and tips on how to survive this sort of thing. They all had Belgian Smidges and spoke their own language, ancient Eridan. Stood or sat in a clump around our chopper, most in its shade and some kept busy by talking or drinking water. All of us were sweating like hogs from the tension and heat.

This layout was simple. To our west lay most of the city. To our north, more burbs, mostly industrial parks and smaller houses. To the south and west, saw wide freeways which crossed on a corner. That defined two sides of this gigantic complex, which was just an empty parking lot about one mile wide, with one huge mall in its center. This mall itself was made of white concrete and was a square block which was 2,000 feet on each side and 200 feet tall. There were a few balconies and windows in some Arabesque design, with pale green tiles in geometric patterns. So even this block had some decor. There was wild foliage covering the roof which was bad sign. This place had clearly been abandoned for centuries.

South and West of the freeways lay nothing but dense selva. I could tell by the tall trees. Right now there was little traffic on these roads. There was one more twist to this plot. In a short note Zeno left on my Apple, said other Terrans on this planet. One called Calpurnia, which was her local name. Real identity was Ms. Audrey Carnegie. She was also from the USA like me. Was happy to hear this. Now lived in Fontana Entheos, some large city further inland. So She was apparently some kind of leader there. She was to keep radio contact with me from there in case I needed advice. The human touch I guess.

"So what's the story?" asked Calpurnia to break this silence between me and my Squad Leader. She was only a human face on a small screen with lousy voice. [This videofone had too much static. How typical.] Chimba frowned; said they had kept this lot clear of weeds and junk like auto wrecks which could give cover to evil beings. There was some litter but otherwise nothing but bare concrete. She pointed at a wide slot on ground level of this mall, in front of us. That was its main entrance. Wore no hat since they tend to be distracting and my hair was by now very short. Still red and curly. We tend to call Aubrey Carnegie by her new name. We could see her on TV. Looked much like me, only decades older. So we chatted for minutes to ease tension. Later, when finally on my visit, me was to learn Calpurnia was shorter than me, yet with more ego. Real crass Earth Ego.

Chimba handed me gas mask, then rubbed down exposed areas of skin with grease in case of Sarin attack. Then we Human noticed once more, how this Anaerobic critter was able to breathe any kind of oxygen based mix of air, despite its makeup. Even so, Chimba had on her belt some small breathing device with Teflon bottle of Freon gas mix inside, in other words, some diffuse combo of flourine, chlorine and nitrogen. Whenever a real Eridan had to exert itself physically, would rely on this supply of **vital gas.** Calpurnia, like most humans, tended to build whole conversations around that one factoid. But We as a military unit, braced with guns and welded like metal into one deadly combo, had to fight for La Cosa. Cosa Nostra. No por heche mierda hablos de Green Haus Gas. Like, the man says, how to pass gas? So finally we all of us compadres feel **ready** to go. Such **Eridan** became cranky. **Natives.** Me could feel it. Tension. **We go!** Aubrey said that she hated these cruel, unhuman outings, even called this a **Gladiator Mentality.** Both of us deep inside felt sorry for ourselves. My bond of Humanity with Her. Zeno had foisted this circumstantial bullshit on us as test of courage and loyalty. That went for all of us draftees. From Terra. As always.

Memory: Being in our National Guard back in Cleveland, Ohio, which was a bit rough but not that dangerous. By the time we had joined there had been no more riots. So this was to be my first real combat experience. Not as some American soldat, but for an alien army on an alien world. Had nasty butterflies in my stomach as bade them God Speed, to start my slow march towards my target. Took me a full five minutes to finally get to that entrance. That immense, blinding wall of concrete loomed up taller and taller as me got closer. Salty liquid poured down all over me. Even into my eyes. Eye Core claimed there were at least fifty **Moloca** in that place. By that mean to say, them **espion** from Zeno who alerted us grunts to hunt down any local Bad People. OK? Simple as that. H.G. Wells called such Morlock. Same idea in my Dystopia.

Entered Lobby which was also in ruins and had no light save that which came in from behind me. In this dim light, could see various things found in most of our own malls. Marble floor, still shiny in places. Fountains and gardens with potted plants. Benches of metal and stone in circles and squares. Pillars that were of marble as well and went at least seventy feet up towards the ceiling. All of this foliage was long dead and dried to husks or rotted to dust from lack of light and water. There were piles of trash all over the place. This central hallway, about one hundred feet in width and 70 feet high, went on until it ended in one gigantic central court which must have been under skylight, for I could see sun beams coming down from high above in the distance. Assumed that two main halls intersected in that court, which meant there had to be four big entrances.

Thus in minutes, had deduced the name of this mall. We think it was called Westgate Commerce Supermall or something, from what Chimba had said earlier. Holding my rifle and carefully moving fore ward almost sneaking, Got nearer to Central Court. To my sides were the usual shops behind glass or curtains. All kinds: clothing, stereo, tools, work stuff for offices, food. What have you? Some of the windows were broken and most of the wares were long gone. Was work of looters centuries ago. This was truly amazing for me. Final proof of how close to the end society on this planet must have been by our Year 1978. Due to extreme automation, there was very little for the average native on Xiotan to do. No wonder the citizens, who relied on basic Work Creation schemes funded by the State, seemed to me like those Eloi.

Therefore, in some places Moloca had taken over. Was not happy with my present task. Still linked by ESP to Chimba and maybe Zeno herself. You see, I had been suspecting for long that such was the case. Zenobia had to be monitoring my mind then. Chimba had the same abilities but was not as strongly engramed to me as Zeno. They did not use radio for now to avoid tipping off our hidden foes. Slowly marched closer to the center as quietly as possible. The critters in here would be easy to defeat since all were Neoform. IQ level 100 at most. They also had to be one head shorter than me. And none had those ESP talents real Eridan casually owned and used, as we knew well by now. Sort of felt sorry for my prey since I had never killed anyone yet in my life. This would be my first try.

Was 80 meters away from Central Court when stopped to remain unseen. Was still in deep shade which was cast by the giant foliage that ringed this big empty space. Out there one vast food court was lit by broken roof windows. There were green fronds of live ferns, palms and other plants. Climbed up into a mezzanine above Ground Level by using some vines and a wide curving stairwell to "recce" this court. It resembled cathedral. High above was geodesic dome well over 100 meters across. Pure sunlight shone down on this scene which impressed me like none other. No where on Earth stood such an edifice. Most glass above was broken, letting in rain. Anyway there was enough moisture for plants to thrive in this hot house for centuries. It went down by levels like an amphitheater. I could its function by many tables and chairs of brass, marble and glass which still lined those levels. Then there were stairs, escalators and signs in some alien script all over. Of course all of it was covered by moss and Epiphytes. Why not Man Eating plants? Am I boring you?

[**Her Conscience.** Not voices in her head. Not Consciousness, as in the new Age. We mean her Conscience. It asks: What are you saying? Ms. Roanoke. Que pasa? Why? Get with it.]

The lowest level was a stagnant pond of clear water with lily pads. How lovely! It was with some droll humor that I figured that all this general effect Jungle as Paradise intended by its Jive ass designers in da foist place? We really need indoor malls in Ohio, which be close to Canada where they really romanticize de tropics. Ah was begin ta' enjoy dis whole episode in some evil way. As lay there behind foliage, scanning do scene thru Feld Glasser, me feel deja vu for my early childhood. Mom in Norte Americano? Random junk in my mind. Bolsa, as rubber ball, thru de hoop. Up on stone circle of exact 12 inches across. On stone wall 20 feet over Court wall, on one side only, of some ancient Olmeca, in Oaxaca? Archeology may tell the Theory, which may be Truth. OK? Now we have seen Miami Vice, as TV show. OK. Then in that show we has Jai Alai as game to be played in South USA. That show was OK by me. They may, in Ancient Times have used Levitation. Or even if so, as Materialist people, did use bats of some kind, like baseball, to make one five inch rubber ball go thru some shitty pain in my fat ass, hoop suspended twenty feet above some Ball Court! Like as me means to say, in the Halls Of The Montezuma, what gives? Physics, eh?

Do Me have, as Mankind, to lick Her ***** Forever, which is nice, as sex, and as groovy Ness, or do Me go up there on some? Tell me more, ass. Am ass. But Me can think. Suddenly spy motion in shallow water of my pond. Wee specks of gold and silver with bubbles. Fish lived here. So Me assume that Moloca had on their own stocked this pond with wild fish. In fact this was my first sign of their presence. So they had to exist. I snapped out of my reverie.

Here is how Alice more or less won her first real battle: She snuck around in random pattern within this maze in pitch darkness. It was nothing but some ancient rooms of concrete which still had trash & bones littering its floor. She used her Sniper Scope apres Nite Vision mounted on her bulky FAL Automatic 200. This fired thirty rounds from a magazine. She also had two Baretta Nine Mill Para Bellums in holsters. The other stuff was nerve gas. That was carried in a gym bag by an assistant. So Alice was not alone. While she advanced in that dark place, other squads loyal to Zenobia were fanning across the entire mall. Alice also used her growing psychic talents. Whenever some enemy was near, one familiar **tiny blue dot** appeared in her left eye. It was energy of some kind. It signaled many possible things such as danger. Thus she combined technology with magic. Since this is not a pure action movie, we will cut it short. Victory.

[Pause.]

After killing my share of Morlock, exited from the top floor to the roof via some stairway and then ended up just standing up there on just another flat concrete expanse. It was yet another standard Sci Fi Golden Age setup if you will permit my true feelings. This scene was as bathed in that same afternoon sun as on any other day on Xiotan. Gradually lost my identity as Anglo North American and so, slipped into my Alien and/or Sumerian one, which I found pleasing to the extent of total self indulgence. After some time spent in a trance I noticed that this place had some shade plus other amenities. Plants grew all over due to high humidity. There were benches of metal and stone as well. Zeno herself emerged from behind some papayas and ficus shrubs, all in a big clump around some ancient rock garden decor. Hanging Garden of Nineveh, like. She was no longer in uniform, basically as officer of the Eridan Army but totally nude. Her dark blue skin was all she wore. Her deep indigo hair had that ancient Sumer Look. Even Barba Streisand would envy that **Look.** And barefoot, in spite of all them broken glass Her citoyen had left on that rooftop centuries ago in some Riot Asino cum party. Party hearty as in ancienne Sumer. Like, in Sumeria as Not as If, Clueless, Such Empress cared! After all, she ruled whole planets and moons, and worried about nada except that ancient strife with her Rival, Imperiatice Ariana. As Contactee me warn Thee! Like as in some chess game, as played by Mark Voonegut, son of Kurt? [**Editor:** That is a bad book to read.]

But me wanna goto up there on Luna, cosa me be Mafia. Am me! Land me on Luna as in, say, the mafia never existed in the mind of the American People. Until 1972. With The Godfather. And Shaft. OK? Revolucion? And no violence? B.S. No way. Was La Violencia Always! In all places on Terra. As Aliens, any Latino, and ancient Roman did call this globe. We must defend it with hydrogen bombs. A some person who was offered Sex, or maybe Love, by some Good Being. As the Being was done, as It was according to Its desires, It. Like, it was Good or Evil, choice made. Was Good. "Me love Earth." It said.

She was, however, to totally indulge herself, armed with another FAL Two Rifle, as I was. That heavy steel and wood thing was okay just slung around her shoulders on a camo cloth strap. She told me that in order to make true progress it was very "necessary" for us all to meet in private, such as on this sunwashed rooftop in my my by now familiar conurbation, Gulf Gity. It was just like being on top of **77 Sunset Strip,** our own code name for **334 Ocean Boulevard.** But what is in a name anyway? Lesbianism was out of the question. How could I ever shove my Terran tongue into her cunt, which was unshaven, since she was made of teflon with body fluids of freon. She exuded sulfur dioxide even now as she stood in front of me in the heat. That was what she sweated out. Finally Zeno spoke.

"The entire thing for us all comes down to some metaphor which is not sexual. I have longed for millions of years for some strong male of some kind hug me in an embrace with his arms of some kind but that passed long ago. We breed by using what you people call Parthenogenesis or whatever. This war of mine and why my own nation needs you, that intense genocidal patriotism, that love we feel for each and every one of us, plus love of my People for their Leader, is none of your concern. So what am I getting at?"

"You know. You were on this roof passively in your Luxor just monitoring me by some method like ESP while I was risking my life down below. So your mind can penetrate whole meters of concrete. So what?" I said in the glare. Both of us, true to our values in all honesty, held deadly weapons but did so casually, locked in tense debate. Zeno continued her story. "This began for you in July of 1978 in a field South East of Cleveland at 4:30 in the morn. We met a Canadian of about your age at the same gas station after he having a meal of coffee and hamburger wrapped in plastic from the same vending machines you have used many times before in your life. Correct?"

"Yes. That one I always used on my drives to and from the South, Kansas, New York State, Pennsylvania, Canada, Midwest and other places we had to cover. In my big, noisy flashy, 2 ton Dodge Fifth Avenue. What of it?"

"Some dude from Montreal. He was about the last of any humans we observed hitching rides over the years from above. He had one final ride from somebody driving some large rig. But important is this: at which place did he stop for one last break before they had to continue on their way? Do you know?" quoth Zeno.

"Easy. It was that place we often see. Have been there many times myself. It's made of concrete and glass, built in 1948. Looks like a flying saucer. Real corny. Or perhaps right out of the Jetsons. We first saw it when I was only two years old as my parents drove me around on camping trips. It was at least 150 feet wide and 50 feet tall. You had to cross some pedestrian bridge of concrete and glass over Interstate 40. I like its design."

"That person was the last one to do that. Probably inspiration for that local myth, the Phantom Hitchhiker. He looked like most of them. In fact he always insisted on having no name. It was as if he were only a ghost."

"Okay but so what?" Zeno told me that my conditioning over the years in that region had made me into her an ideal subject. There were just two blobs of light in that area as seen from above. The restaurant which looked like a flying saucer, plus that isolated Esso station with its one lonely employee only 20 miles away. They had been busy monitoring that part of North America since 1947 which was no big deal. Watching us as we came and went. Men, women and children.

Watchers were often Omgal agents, or perhaps the tall lizards of Cetiwana. But usually Eridan. For their wars it was convenient to choose me at random. They had to catch some person so high on her THC habit that her ESP was cancelled out. I was actually stoned on grass that morning. Insulated my brain against emotion, aka vibes. Which was good for me and Zenobia. The ship itself looked exactly like in those photos from Los Angeles of 1965. The famous Heflin Polaroids of August 3, 1965. Zeno told me It was just container of brass, aluminum and titanium alloy which had no engine but was teleported hither and yon; over Fenn and Dale as it were. The thing had hovered in stasis over that phone booth for one nanosecond or whatever, then teleported me inside. Then it vanished, to manifest in some larger ship above our ionosphere.

Zeno then explained why she had appeared in the "raw" to me. At first it was not sexual by intent, although I had already surmised what her race was like. Without any males, totally Lesbian. Merely said that we may as well see just how alien her kind really was. In general the Eridan as the most advanced Archaea, were more evolved and complex in every way than viral microbes. Archaean beings throughout our galaxy had only 17 strands to their DNA while we Eukarya had 23 as far as I knew. By the way, most of this genetic science had been given to me via my floppy discs on my quaint Apple back in my suite at 77 Sunset Strip. One of Zeno's specialists had urged me to study basic Genetics. Zeno stood there naked on that hot roof in the glare of her sun and lectured on. I grew tired and sat down. "We are billions of years old, mon amigo. If the **Steady State Theory** holds, then my race may be tens of billions of years old. Yet if that Big Bang theory of your Terran science holds true, then only two giga years, maybe 1.5. But that still gives my Eridan far more evolution time than your own mere **three million.** Question?"

"Did your race really originate here on the second planet of this star, of many in our galaxy? Are you sure?"

"No way. It could even be were brought here trillions of years ago by some utterly lost beings or forces from anywhere. Like other galaxies. However, our own historians suspect strongly that microbes made of teflon evolved here in primordial soup made of very dilute hydro flouric acid, with plenty of sulfur compounds for food. Add some tasty carbon to this mix, and we grew quickly into what you see in front of you. In fact, in a biome of 300 degrees Celsius, that was easy. Faster than you Terrans in your oxygen-based Biome in 3.8 giga years. It was easy."

"What was it like?" I asked. Zeno sat down while sweating sulfur dioxide and continued to describe her world. In ancient days, it was one with slightly lighter gravity. Vast but shallow oceans of very mild hydro flouric acid covered most of Xiotan. Land covered most of this and was flat. The boiling point of local liquids was kept higher than it should have been by our high atmospheric pressure. So their oceans were still liquid and mostly H2O. Silicates, some metals and massive amounts of salt made up the crust, which was poor in iron. Glass and pure metals could not exist under acid rain. So most rocks were crystal layers of sodium chloride and what not. Thus the smelting of metal was not ever invented by us Eridan, who were sentient but

forever in the Stone Age. They had to use only brittle crystal tools. They could never achieve what we Terrans had in so short a time had achieved. Their culture was stunted. Obviously space travel was out of the question. She went on in detail about the geology of Xiotan. For example it was always 7500 miles in diameter with a molten core of iron and nickel. There had always been volcanism and the planet was much like your Jovian moon Io in many surface. In fact, even by 1978 over one billion Eridan of her Tribe lived on that distant moon of yours. After all, that biome was congenial to us even without an atmosphere. Having sulfur to eat was the point. But that was Io. Our Home World itself did have a dense atmosphere of halide gas dilutes with argon and helium. Land was covered by flora that was strange as Hell to you Terrans. There were things which resembled lichen, moss. Algae had not yet come to be. There was seaweed but of wildly differing chemical makeup. On land they had Archaean versions of grass, bushes, herbs and shrubs. Also ferns and trees.

Our highest forms were things like conifers, which actually had solid trunks and branched out to form a canopy. Yet the tallest of all were 80 feet at most. Gravity was strong enough to work on these simpler forms even though wind was weak. Leaves and fronds had a reddish tint, going from dark orange in case of lush foliage full of sulfur hydroxide to blood red in leaves of taller trees, which were exposed to the best sunlight. By the way, the sun was just as bright as it is now, which is blinding pure white. "As you can see by now Alice..." she lectured on. I was totally absorbed by her account. The undergrowth tended to be a weak pink or sick violet color. The basic point was that Vitamin A was in this primitive biome used for photo synthesis. There were also animals in both water and on land. Most were still halfway between flora and fauna, so that it was hard to tell diff on this planet. "And what of your culture?" I asked. Zeno said that there was not much to tell. They were tribal hunters for a very short time. In all life there was no sexual act, since they just budded off like microbes. Even sentient forms, Her Race, had only one gender which was only female. Only "Y" chromosome without any "X" which wasn't needed. Only the highest of all forms here took on mammal form such as an attractive female body with narrow waist, wide hips and two boobs. Her milk even resembled our own milk. Thick yellow cream based on water and sodium, but rich in sulfur. That one comment made me laugh.

Then she stated that Eridan invented houses, fields and the usual stuff fast. From Neolithic to agriculture. Then came language, math and astronomy.

Also had some simple alphabet like our own. Again back to that lack of iron. Since the crust lacked iron the only metals were soft ones like copper, aluminum, tin, zinc and the like. Stuff like lead, gold and uranium was down below due to its greater mass, so any fission or fusion technology was impossible. No gunpowder nor other explosives were in any way developed. The best they could do was use fire for smelting, but the only result was some crystaline alloy of metals like copper combined with flourine and many other elements. After all, only copper was used in their blood for valent action. It was vital for breathing.

Instead of "hemo" they had cuproglobin. They ended up living in great cities built of crystaline stuctures. Pyramids, hirises, roads and all of any advanced urbanized society. They even had an Empress ruling her world as respected leader. But of course they could not physically go beyond whatever Sumeria or Mayans had. No way did their industry go beyond about 4,000 AD on Terra. I asked suddenly about telescopes and really advanced astronomy. Zeno paused. Then she explained that, like Central American Indios, or Incas or even your Nazca culture, they had advanced concepts in science, but only as concepts. Kept on records. Not any technology such as our own. Pure smelted metals, wire circuits, gunpowder, engines of any kind, also nuclear fusion in any way. It was easy for my ancestors of long ago, say 900 million years ago, to create or discover basic principles of any science and record them. But we had to try harder to make tools. She gave an example.

Here it is then. Yes, I'm still your Alice. So they knew concepts of astronomy. Somehow. By intuition. Having ten fingers and the same brain as our Cro Magnon they created from nothing our native system of Math. As that is the only totally abstract science in the universe. We had the same Decimal system as you did. And do. We did not, however, also develop an alternate system of Twelves. Dozens. Like for egg cartons or your Zodiac. Your Zodiac with twelve signs is native only to your own star system, with its basic pattern of twelves. At least until you hit Jupiter. At this point had to tell Her that me personally have no talent for Math or Astronomy. Really. "Do not worry. You need none. Like, we can take your brain and pump vast amounts of cosmic data, known only to us aliens, into it. You can be dumb fuck, for all we care. Make you look as genius, dude." Me was shocked! But would it not destroy my brain? Zeno stared at me. Tried not to stare at her bushy indigo teflon fibre fotz.

"It will drive you nuts but you are anyway." She carried on with her lecture. Her scholars made telescopes of crystal which allowed them to study the firmament as we did. That came to pass long before our Cambrian Era. We only had telescopes by 1608. As soon as the first Omgal craft landed in 800 million B.C. they handed us technology we had need of. All of it. As in, nukes, guns, computers, space travel. Anything. And then, being warrior nation, we conquered this galaxy. Many Eridan empires exist within Milky Way. One more detail: This planet has no seasons. Yet our moon Safir revolves around us ten times per year. Ergo ten months per year. Again we have on Xiotan pure and locally evolved Decimal System. Even the Omgal use it. Okay? We stood in silence. Then I had to sit down to relax. As joke, Zeno told me about some metallurgist from Essen who could explain this to me. "Essen" I said "As in to **eat** in deutsch?" **Essen.**

"No. Just some city in Europe. We will eventually return you to Terra so you can meet Cindi Nord. She shall be your assistant. You can look her up in our database. Consult Army of Zenobia, simply Online. Basta." What a bizarre culture! I suddenly had a real spooky thought. This idea was complex enough to be avoided by our own Ufology cults. Those experts soi disant. It had never been stated in any of our standard scifi books. It was this: Why rely on ships to traverse the Cosmos? Then, as soon as Zeno had succeeded in fully explaining her culture I had an almost physical revelation. A satori. If their masters could teleport at will, then did they not in short order make it to other worlds? How about their moon? The rest of their system perhaps? Did they not just colonize this entire galaxy plus even others? Their path to the stars was faster than our own. Zeno agreed. It was so and I had guessed correct. Good! Twas what she wanted. Then I asked about the rest of that equation. Basically they had colonized most of our galaxy and had vast empires with trillions of Eridan living on many worlds. They did not get ships until 800 million years ago. They got them from the Omgalii who were not familiar to me. That Race was only 800 million years old and came from somewhere near the center of Milky Way just "recently" and had given them FTL ships plus atomic weapons for nothing. They needed no actual "goods" nor "services" for all of this stuff. Zeno paused at that point in her lecture.

As an American I figured out the rest. It was easy. You see, these Eridan and their silly worlds had nothing a real Star Wars kinda race would ever need. Why did the ancient Omgalii, who were also warriors, but had superior industrial power, do anything for those essentially Neolithic Eridan? It was fear. Some Omgal gurus could master cosmic skills. Such as I did. Yet Eridan were much better at this. They had the option of appearing on board any Omgal ship. Then steal it away. Killing the crew perhaps. Like on Star Trek. Being smarter still, Omgal leaders decided to trade with Zeno's ancestors. No war. They fought each other anyway. All the Omgalii ever got in return for machinery was currency. It was often on paper. Not only that but shares in stock markets, which both races also had anyway. So all advanced races had this in common: the value of anything was based on nothing. Totally arbitrary. There was no such thing as a "galactic" credit unit. That was one idea Zeno assured me of as the truth. She was very serious about that point. She shot a hostile stare at me right then. I looked away and spent some time pondering this.

Yet she was right. In 1913 my own Federal Reserve Bank deemed all US currency to be legal tender without benefit of gold. All that shiny metal in Fort Knox had only symbolic meaning. One of the James Bond film was merde, since it did not matter what happened to gold in the USA. In fact, by 1933, parts of Europe had banned the gold standard. It was the Socialist and Fascist regimes in 1917 and later on who deemed that money was to be based on real production by the real labor of men and women. In Russia and Germany most of all. This may lead to massive genocide. By 1948 this global society had been fully explained by Orwell. I based my thesis on it in after my B.A. His novel was also used as text in Polisci 350. I had study to make it in my field. We assume aliens have similar motives. Zeno ended her lecture by promising to introduce me to an Omgal trader. He was an ally of hers.

She dressed in her uniform again, that familiar sexy bikini which glittered of titanium in the sun. It was so typical of her to have to point out her Thingness to me. As we waited to be airlifted out of that warzone I imagined her as a savage Eridan native, wandering alone in some jungle of goofy plants under her yellow sky in an acid haze. Hydroflouric. Naked as she waded across some shallow pond. Perhaps she carried a sharp crystal tool like spear and with dagger on a belt. She might have been hunting. That was a private vision I could get into. A few smutty Sapphic ideas floated about as well. This is why all of us enjoyed Golden Age scifi so much. Sex and violence sell. It just goes to show that personal identity can be hidden and romanticized endlessly. Such as look how we can hide crude ideas of commonality. They arise before puberty and also exist on this world. By the way, I assumed that this new person called "Flavius" must be male. And of course breathe oxygen. I got that name from her mind directly by picking it. Her verbal speech was for effect mainly. We slowly came back to reality as her squad came up the stairs to evacuate that place. A dozen prisoners in chains marched across parking lot far below. More grunts arrived with armored cars to carry them away. It was all vague and devoid of meaning. Zeno did very little. The others did some yelling and fussing with each other and their gear. Twenty Chinook choppers slowly came towards us to land on our gigantic roof spewing up dust. Loud as Hell.

We quickly climbed aboard and lifted off. As for Ariana who had been vaguely hinted at, they did not show up. For some reason our "real" enemy, whatever they were, never showed up. Was it on purpose? Were these advanced beings more or less deliberate in their routine dumbness? This ambient hate. It was all one big joke to Zeno and every one of the others. These Morlock who were just pathetic victims themselves. They were all Neoforms. Natives. Even most of Zeno's army, as far as I can see, are natives of Xiotan who need oxygen. Slaves. What about racism? Maybe I smelled a rat. Had this nasty feeling that none of Her prisoners were to live long. Their flesh may be served up next week in some super mart. Who knows? Ask me if I care. Wonder what Hugo Gernsbach or J. Campbell had to say about this. As you can guess, I questioned motives of my new comrades in arms. Golden Age, eh?

[NOTES ON SCIENCE]

[**Author's note:** We are unsure of what the exact workable or "feasible" kind of temperature this "Archaean" biome might be. Teflon may be pliable at this high temperature, since it is known that this compound melts at 250 Celsius Above. So in our chapter above, in some passage wherein Galaxa finally tells our human "hero" in plain English where, what & how her "race" came from. The field of Exobiology has to be about 150 years old by now, which means H.G. Wells invented it. That lesser field of "Alien" Biology started in about the 1940s. By that time, some very good fact & fiction had been written dealing in creatures which were either Halide or Silicone compounds, which we state here in a very rough way. We do not hold any Hard Science degree. Okay?]

We shall also add my own discovery, some Being that could drink Cyanide diluted in water (weak but still strong enough to kill any of us) and also contain this poison chemical within its body fluids. This goes far beyond what any schizo can do. Their skin color just has to be blue! We actually met & talked to such a Being once in a large North American city. This being seemed to be female but without tits. It was even sexually attractive but underage. Maybe we can call them Cyanophiles but They may not be for real. As you may guess by now, this being may have been some Early 1990s freak with black clothing & blue body paint. Anyway; to return to our point, we may have to lower the basic temperature of their Home Planet, their Biome, somewhat. Or more? We feel strongly that water is the key here. At 100 Celsius all water becomes gas. Since the body fluids of such aliens contain sulfur and freons, all of which are reactive as well, they must still be dissolved in water of some kind. So, unless atmospheric pressure is far greater than here on Earth, climes must be far lower. Off hand, we may guess, about Minus 100 to 40 Celsius. That may satisfy you Purists from Chem 301 and Physics 201. Fen, etc. A bit of Genre sarcasm. As another joke why not rename Biology as Triology for having three branches? See relevant data Online.

END OF CHAPTER FIVE

Next: Chapter Six

Seamoria ⟨⟁⟩ ∈
aka Cimora
President of Planetary Council of
Premiers of Xiotan in 1978
Eridan Neoform by birth in 1928

"Fontana Entheos"

After one night I settled down to my routine as just another human on Xiotan. Now keep in mind that **we** have already seen **Ms. Carnegie**, thus had to conclude that some other Terrans were held hostage here somehow. Yet I did not get to meet her, aka **Queen Calpurnia** in the flesh until long after my Mall Battle. Such was of no interest to me. Was more into meeting this **Flavius** critter who was supposed to belong to our famous **Omgal Race.** He had to be far more vital in terms of strategy and politics to Zeno and other Xuan. Was tired just after getting home after our battle in that shitty mall, so I fell asleep at once. Next morning at Six promptly, my alarm clock woke me up. Had by now various digital devices which were supplied by the aliens. They resembled our own LCD clocks and even wrist watches, which by 1980 came to be more in use on Terra and were designed with us in mind, which shows you how any gap in cultural levels was closing. Had even become, in my own way, an Eighties Person.

At least this cultural cum techno gap of mine was closing on Planet Here and Now. Had by that point been living on Xiotan for five months busy analyzing alien society. On my second day, suddenly noticed glimpses of what must be snow covered mountains in the distance from the mall roof. At first I dismissed them as being just cloud banks, but that was not so. About 100 klicks away or so there had to be glaciers or something, which were covered with powder so white it hurt my eyes. Suddenly some random unit of mentation crossed my mind. Like, all but one of my timepieces here in my suite were on **their** 20 Hour Day local system. My only watch with **our own 24 Hour Day system** was, well, my Swiss piece with its gold plate, given to me by some relative.

That short flash stopped my train of thought, focused as it was, on this unknown **Flavius** person. Some small furry Thing. Fuck him **or It**, my mind went. Hmm… was this critter supposed to be divine as well? Like his amigos? Ha! Maybe consult my database? Fuck that, better to concentrate on my other impulse about that chain of high mountains. In this heat they had to be at least 15,000 feet high. About like the Andes. Consulted maps on my Apple; came up with good ones. In color even, which must surprise you. No PC in those days had GIF. Only DOS. Green letters no smaller than 36 picas on black screen. Yet my hosts had already created some new Graphic Interface for me as luxury. Was years ahead, far beyond the Eighties in concept. Why? So ask Flavius, then! He should know.

We had maps with full color in very high resolution. Up to local standards anyway. There really was some mountain chain stretched along a thin line running from north to south 100 kilometers West of this urban area, or at least somewhere beyond a wide band of thick selva with which lay just beyond the outer suburbs of Gulf City. By the way, according to this map, this conurbation as a whole was one blob about 70 miles wide. This was built over the swampy delta of some large river. Name unknown. Must be longer than our Amazon. It drained some wide tropical floodplain which looked like any on Terra. That river emptied into a large feature called the **Great Gulf** of what my map called simply **The Great Continent.** Wow. That was thrilling, to be sarcastic. Except it was an ancient bomb crater. To the South of us were flat plains going on for hundreds of miles. On each side of this Great Continent was one long mountain chain, and the western one was labelled Empire Range. Some town called Fontana Entheos lay on an altiplano in a lower part of this range. We made note of this fact.

[Night passed.]

Sound of alarm going off. Was only my Westinghouse clock radio. I noticed it was close to dawn. Twilight. At risk of boring you, it was fact 9:35 in the morning, but we had no "AM" nor "PM" here. This planet only has 20 hour days. Also it has no seasons. Each day has monotonous 10 hours of dark alternating with yes - exactly 10 hours of light. All fucking year. We are almost on the Equator here; make that 8.5 degrees of an even 100 miles North of it. The other large cities on this continent are **Tapiran & Fontana Entheos.**

Had my bath while dawn came on. As I sat here in my tub with door open a warm, lovely red ball rose up in the East then went up the Doppler Scale to white hot in minutes. This was so awesome! Later, sat on floor naked while having coffee. Being a bit absent minded, gathered wool. In a brown study. Suddenly near the bottom of my cuppa realized that none of us had seen Etiana for weeks. With shock that moment, realized how socially isolated all of us were. We tended just to spend whole days or even weeks alone. Inside, or out wandering in some jungle. Reading, watching TV, meditating, etc. Or commune via some social media. **But Etiana must be dead!** I had not even spent time thinking about her. She was not on my agenda. Galaxa had been speaking to me over phone mostly. She had complicated plan for introducing me in person to some aliens who also lived on this continent. There lay my rub. Zeno had been making me take notes on paper and modern hard drives which had 500 giga bytes of capacity. Again, wow! **She was getting me on track with my hitech stuff.** My PC was now a genius. Also my friend. Now Zeno or her staff tended to shoot masses of data at me every day. I was to learn how to teleport eventually. She was to tell me of her plans early today. It was on my duty roster. I felt like saying, "Zeno, stop dumping these databases on me! Speak to me! Please, you whore, you Sapphic Ass! treat me like a human being!"

Then she might yell at me from skies above like goddess she truly is: "But I am not human! You are. So do as I say already!" Shocked while standing nude on my kitchen floor wondering about breakfast. Yes, some evil voice of a woman came out of one vent above. Ceiling. Her Imperial Highness encore. "Turn on your HAL-9000 Goofus! I gots to speak at you." This was silly. Never got over how childish & hostile aliens tend to be. This was a land of contrasts. As they used to say in National Geographic or some nerdy coffee table book.) So I did as told. What was I supposed to do? This was Day Two since our first raid on that gang of Renegades. Just a minor part of Zeno Galaxa's War Effort. To "save" this swampy world from some evil enemy from some other star system. Now do we understand? Stellar politics must seem complex. Hope I've explained it well this time. Digression. So I put on some gear and turned on my Apple. It was now upgraded. Zeno appeared. (What is it now?) She was herself, yet calm. In a sort of relaxed monotone a blue face on my small screen intoned "Buenos dias. We wish for you to risk your life again. But your rewards are big. What do I have on my mind? Just allow me to teleport you over to a city some miles away on an altiplano. Are we hip?"

"Hip we are. Stop imitating Wicked Wanda."

"Alice. Who cares? None of my slaves read Playboy. Have no sense of humor. We don't care about your boring society. Now get ready to make a jump to Fontana Entheos. Your man **Alfie Bester** called it to **'jape'** in his 1956 book **The Stars Are My Destination.** Now do you dig it? Get hip to get ready. It is only one jump but you need to bring radio, weapons, tape recorder. Also be in uniform. You got any questions?"

"Who do I meet?" Zeno named Calpurnia plus some strange furry Thing called **Trader Flavius.** And was that a Latin name? More aliens again.

So anyway, as events turned out, I had one hour to get ready so that made my Jump Time 12:00 Noon sharp. Also my ETA. It took less than a second. Somehow I recalled not having read any Bester. Zeno is a sport. After committing mass murder she walks out naked in public just to prove an old Earth saying: An Empress wears no clothes. I assume her crew was used to this bullshit. So why rub it in? They did not even laugh at me in sadism. No emotion at all. Aliens! See my point now? I'm still not used to this wild planet. Their society, as Mr. Wells might have said, "Has gone past the zenith of decay to its nadir." Or something. Then I stood goofish in the center of my room ready in full combat gear with my FAL Deuce, Parabellum plus other useful stuff. Zeno was gonna get goose bumps! Eridan was burning up some more hydrogen atoms at its position in the deep azure skies above these sylvan shores as I saw Her at Her zenith. High Noon. Like in a cowboy movie. Me like an eight year old playing War.

[**Interlude.** Now go get some popcorn & pop!]

Editor's note: We know this sounds cheesy but - hey! If you think this stuff is phony or a pain in the ass let us know. Look, I know... but consider that H.G. Wells or Jules Verne could not get into randy stories like this one. Those were Victorian social mores at work. You are simply lucky that Reaganites might be at work here. Otherwise you might have read some smutty Rap lyrics mixed in with Cyberpunk and/or Urban Mythology. Don't throw stuff at me as if I were a poor hippo in the zoo. Just imagine being really abducted by aliens. Like what if the "Tran" series by Jerry Pournelle were actually true! That could easily be. You see, what we figure is that Tran revolves around Alpha Centauri with Beta as one big additional yellow giant throwing more heat on that Medieval fantasy world. Meanwhile, just say that your Demon Star must actually be Proxima Centauri as white dwarf which every 600 years come close to Tran on its fucked up orbit to really fuck it up heatwise! So mote it be. But what if Mr. Pournelle is even right in some ESP/cosmic way? Who knows? It's only fiction.

[**WC break over.** Now walk back to your seats.]

Zap! It was a Jape of sheer Beauty & Joy! As Bester may have put it back in the Fifties. Jet Age turning into Space Age. We call it "jump" more in our prosaic fashion. Well, after Con manner. So Zeno did finally teleport Alice a few hundred miles into... another building. Alice felt sweaty & overheated. This feat always pumped huge amounts of energy into her body. Brine at fever levels poured out. She knew well she was able to handle core body temperatures of far lower than 25 or far above 37 Celsius with 32 being the norm. So she was okay with fever. That made her also resistant to most microbes. Such was her basic thinking. She shook like a leaf. As she fell down upon some part of the floor of this room, another concept came on as comfort: Like after five months in such a warped society of aliens, how can that silly notion of some of us having altered genes leading to biochemical unbalance be valid? That would be fascist in itself. Freud would not agree. Then even as Seventies and Eighties Urban Myth progressed, these 1% of our society might even be Cosmic Supermen.

Maybe Androids with amazing bodies, like Rutger Hauer in Blade Runner.

Alice was still in the Year 1978 and that movie had not been made yet. She was not even a PKD fan. In those days they still had some rudimentary notions of Democracy down in North America and Western Europe. Was all she knew. Alice Roanoke. Anchor Wat? Fuck these jokes. I mean what of Anti Psychotic drugs? Were they just downers to take the edge off that Adrenal fluid in the blood of apparent **Schizophrenics,** creating **Android genome,** which was like supposed to make them into **Nazi Freaks?** Miracle drug, they say. **Thorazine.** But we digress. Thorazine was known to Alice who was streetwise. Hip to most drugs legal and illegal. But all those miracle drugs were created only after 1948. That was long after Freud and Einstein. But also Marx, Lenin, Darwin and Nietzsche with their atheist concepts which might have helped us. Nor did she know Zeno was to take her into Earth of the Year 2084, but we do. Why not get hip to creatures from beyond Time and Space? She came out of Her trance.

Alice slowly rose up. Had lain some moments on a marble floor just to rest. Still perspiring. Taste of salt. Zeno stood in front of her. She did not react. Just steady gaze. "So did I do that?" Alice asked in an even voice. This had been better than before. "I did that." Zeno said, "You are not ready yet. It will come with practice. Tat tvam asi." They stood within some vast space inside. It was dark. This was like an airplane hangar done in salmon colored marble with striations of brown. Thick stone pillars held up dome. Was pieces of furniture along distant walls. Gilded wooden couches, beds and tables with plenty of soft cushions. There were huge brass chandeliers hung from above plus some giant candlesticks on floor but none were lit. What poor light there was shone in from a few open windows. Alice was glad to be inside from constant boiling heat. Cool breezes blew in. There was nobody here but Zeno & herself.

So her amigo had teleported her here. Alice wandered about in circles then sat on some table. How rude? Do they sit on table tops here? Or is it not kosher? After all, this table was only two feet tall but 15 long by 10 wide. Was it created for some kind of wee folk? Hmm. Finally Zeno said "Are you trying to polish that table with your ass? It's shiny enough as is. I just said that for comic relief." Alice gave her Empress mentor weird look. This reminded her of the Hagia Sophia. She had very little to say. [**Editor:** Hey, jerk! This has to be some high school jerk thing where some chick with a skirt with nothing under that, plants her ass on some desk, or even worse, some poor xerox machine, and has her bare Arsch sit upon some cold surface made of wood or glass. Right? Is this Thrills Ville? Like, do we on Earth do such to expose our vulnerability? To being fucked?]

[**Tomas:** No idea. Me nerdsville. OK?]

[**Senior Editor in N.Y.C:** Kindly resume our story. OK?]

"We are in luck. They are not here yet. Later on today Queen Calpurnia and her great court will fill this room. Seems to be empty. No food on table. Must be busy in her office which is in this palace. Come. Shall take you up to your designated room. It's only one of many Guest Rooms provided. My slaves ordered this done to mine own satisfaction. **My Empire pays.** Okay? You are tired so we shall take an elevator." Suddenly, in silence, Zeno walked away from me. We somehow slipped from one narrative style to another. Yes, it's me. Not any editor but you know, I am beginning to notice how this tale moves from of one persona to another! In some video games, which me like to play, this is done by pushing a certain button.

Loved it back on Terra when in some penny arcade or on my Gameboy from Atari one can move from some third person role to my own first person in my narrative. Okay? So fuck you & all your ancient mores. We just dig how we can move from one body into another or perhaps take on the point of view of another Being like after some better Satori. Zeno was in some other room. Was finally able to relax after another alien action upon my person. My Master was sure to come back in again to bug me some more. Yet at this moment showed **presence of mind.** Carefully pulled blanket off some bed, laid that on my table, then placed my guns and other stuff upon that. Also my back pack. Had brung along lots of ammo, water & food.

Heat. Still day lite out. Alice got up from her position of sleep on cushions and blankets, all with some **Levantine** patterns alluding to **obscure ancient cults.** Some alien voice came from her radio waking her up. She must have slept for hours as it was 4:30 PM on her own watch, Rollex from Ohio, yet only 13:45 Eridan Standard Time on her **native** watch. The latter was an oval LCD display on a golden metal band which covered most of her right forearm like some Roman anti sword **wrist guard.** She loved the sheer erotic feel of it. The numerals were in Sumerian Cuneiform of glowing yellow on black area. She woke some more and wiped her own eyes. Was still dark inside. A few dark figures came in from shadows. Zeno was first. She came in wearing her usual: Shorts, halter top & boots of silver lame. Also bands on her throat & wrists of titanium alloy.

Smiling she asked "Want some lites?" Just to be sure, Alice said "No" since her eyes had adjusted to low light. Who needed to be sucked up to? For some reason; within this whole setup, never had Alice been challenged personally. Beings on this planet tended to treat each other as equals in practice whenever they actually seemed to be on the same **side** in some social conflict. Otherwise, would personal reality even Add Up? To anything? No. So much for Orwellian fantasies, in case they make you horny. It helps to **read** it. Blowhard. Zeno agreed with nod and waved to indicate two beings to her left. "Introducing Calpurnia & Flavius... uhm. This is the Terran Me has been talking about. Okay?" Alice said "I have already spoken to you on phone. With video so me now recognize your face. Pleased to meet you. Zeno?"

"Likewise me be sure. Speak English, like Zeno. Totally. Please call me **Audrey Carnegie.** Was born San Antonio, 1930. Later in 1947, was kidnapped by some kinda flying saucer that brung me over here to Xiotan. Look. Lady, we really do not understand this whole bit. Much less than you can! Ya know why? Huh?"

"Dig it. You're askin'. You look only seventeen! Yet you should be 48. That means you were taken through some kinda relativistic continuum directly into 1978. Was that so?" **Alice was surprised at how easy it was to have a serious conversation.** These people did not seem to be into comedy but at least they were more open. Zeno explained "Here it is: You only had a short time on fone. An intro. Just to display courtesy. Never mind her local name. It was given to her by the Sirians who only speak Latin. So they named her after some lady in Shakespeare play. Julius Cee. Anyway fuck that. I speak fluent Sumerian & English. That's only 'cos I lived in ancient Chaldea as a goddess for a few baktun so I had to learn their lingo. Can even write cuneiform. Done learn English one century ago yet can dig your mod talk, even. Any questions?" Alice ignored the girl who stood to Zeno's left. A fellow human from Earth. So what?

She focused on a third party. This creature was only three feet tall and silent. Two large round eyes stared back at her. Alice felt creepy vibes. It was covered with thick white fur like polar bear. At least this thing was bipedal. Nostrils hidden under fur. Under its eyes was a wide mouth. At least they resembled our own eyes, with yellow irii. Hmm? Irisus? [Okay. In Latin, or medical lingo, be it that circle around our Terran pupils. Call it the Iris. We all have only two. Then rest of eyeball be white with blood veins. Okay? At least on some Man from Earth. Right? Just wanna be sure.] This critter wore nothing. Just some kinda pygmy bear. After a while Alice felt it was cute. For a short while, its furry aspect reminded the average human of spiders. Then opens mouth to say **"Me can speak English but not good."** Zeno was annoyed with this dorky bear. So he has lied to us all, She figured. **Some ally! Me like Donna de Terra much better! Honest answer.**

[**Note:** Calpurnia was just another cute kind effort at humor from Fluffo Flavius, that alien Alice had just met. Both of them spoke spoke English. By the way, Trader Flavius was of male gender. His kind have the same sexual habits mammals from Earth have. For some reason, Eridan tended to speak Sumerian while Omgal, that Race Flavius belonged to, stuck to Latin. But as of our Elizabethan Age, both races also learned to speak English as Terran lingua franca. Keep in mind only a few aliens even bothered to learn any of our languages. What Flavius & his Omgalian race are and intend will be explained later. Now for description. Audrey Carnegie was as we know an American. She grew up in Texas in 1930s in high school but also worked in her father's general store. They were a

WASP family like the Roanokes. Audrey was brunette and short. Like Alice she had full figure. Call her an Eleven. She used to wear, like Alice, jeans & boots. **Like usual. As before.]**

Unlike our hero, she had ridden on horses. Of course she came from Low Tech kind of society. Unlike Alice, she had openly laughed at anything **any** aliens had to say. She even broke up at Zeno Herself, loudly calling her "Some two bit jackass ****** whore" due to her color. Blue. She at once was ecstatic at their muggy Amazonian clime which reminded her of the Deep South. No kidding. She also demanded to be handed **pieces** such as Smith & Wesson 38's and Winchesters. Even better so. In other words, was almost a cartoon character. Alice had a veneer of culture but this woman was savage. En total! Yet after Flavius had set her up with power her estupido act vanished. She got straight. Then adapted. Was even more of a natural to Alien Life than any modern Earthian. So perhaps Calpurnia's heritage was some how alien after all? Becoming Queen of an entire planet helped.

[**From any UFO Network:** Calpurnia had asked her captors at once about a famous **case** from 1897. On some Kansas farm. A farmer had seen Something that of course in retrospect had turned out to be some German Zeppelin which was on some spy mission over America. No, Miss Carnegie said "It was not Martians nor we feel any other aliens!" And so on. She was never fooled. She even right away got into quantum computers. Using them she made herself into the only local Terran migrant with Masters in Cosmology. She even improvised on basic Wave Mechanics of Hendrik Antoon Lorentz. Took to alien society like fish to water. Her dress was savage.]

Unless it was on some formal occasion, she tended to be topless. This was due to heat but also her innate lack of taste. Most days, she wore flimsy white cotton mini held up by a heavy leather belt round her waist. To that was strapped her cowboy revolver on ammo belt. Lots of jewelry plus bead necklaces but these failed to cover her tits. Her pale white face was that of a pretty teenager. Hair short and red. Wore sandals on her feet. Basically same style as some courtesan from an ancient Mynoan kingdoms on Kreta. In fact aliens accepted her totally. For some strange reason Alice suddenly came to realize that ancient cultures must be inspired by ideas from the far future. What made her think so?

[**End of notes.**]

Back to reality. "Hey, wake up." Alice did so. She had been sitting there in a trance wondering about these two. Zeno pointed out that Audrey as Queen Calpurnia was the only Supreme Ruler of this planet. Then added "nominal". As if that was not clear. Yet She was still, as Empress, Ruler of the entire System. Okay so far? All three stared at Alice. In a weird mood she just replied "What I get from this encounter is a sense of wheels within wheels...." Then she was at loss for words. Calpurnia was now in her Royal Uniform. One Roman toga over her usual garb. She strode over to another table covered in office machines and the like, then picked up an 1800s model rifle. Winchester. Hidden by Her classic garb, as if Caesar Himself presiding over His ancient Senate. There was also a Six Gun strapped over her hips. Then came back. Aiming her heavy repeater at Alice she said "You are my own trooper. I jest got here back in March of '78. So me was here long afore you. Got the jump on you, which is all that matters. In this practical, very physical universe. Right?" Alice mumbled, as if speaking to a funny white rabbit "What are you babbling about?"

"How naive. So we babble? More like Babylon and you are no Flash Gordon. I seen them swell serials. Me, I own this fucking joint. This entire planet is mine! Zeno & Fluffo can allow me to blow your sorry noggin off. Now what?"

"What?" asked Alice. The very sight of this evil being shoving Her heavy Winchester into her gut stopped her mind. She even felt like asking "Are we on 'F' Troop here?" but controlled her tongue. Being smartass leads to death here. Simply asked what the "issue" was. "I am too abstract" she decided. Better listen to advice. Take orders. Miss Carnegie chuckled and said "There's a good book for you my friend. Not Bible. Actually, I like that but better for you is How To Make Friends and Influence People. By another Carnegie. Read it." Smiled, lowered her rifle and walked away. After her nemesis had receded into shadows, Alice pulled cushions off table and relaxed on them. This palace belonged to Calpurnia. So did that city outside, and so forth. So this is how we make things "happen" here. Zeno said to all concerned "This is not my way. It is hers. You two Terrans figure it out. Cal will accept you if you agree to our basic plan, which we must explain to you later today. As for me and Fluffo, none of us aliens wish to become violent. That is your thing. You & your kind."

"You are now my little amigo." said Calpurnia. "Yet we can tell more. Amigo was almost used as English word where I come from." She was speaking from behind curtains sounding Phantom of the Opera. Why not? This was turning into an epic by Cecil B. de Mille. Why not copy their style? "I shall suck up to you. Yet this is no joke. Yes, I am very much afraid. Maybe between us two Proposition Thirteen? Like no taxes. No more evil socialism..." but Cal said "That was stupid. We are no longer American nor Terran, nor fuck all in this place. You belong to us. Now get with it." Patronize none, that meant. When on Xiotan, do as they do.

"Do as we say" said Zeno, "We shall explain after. Have some food for after this nasty drama we must feel hungry. We have nice dinner cooked for you at 15:00 including vino, beer and dope. Luxury for prisoners." Zeno sat down on that table while Flavius made itself comfy on cushion. Seems to be perverse alien custom. Cal clapped hands to make servants enter with platters of warm food and many drinks. All were young females of many Races. Was this by Cecil B. de Mille? "We are treated like royalty here. And yes, we all speak English. Like we really parlais d'anglais!" came from her Highness Calpurnia with joy. As dinner was served their servants danced about like Tantric Yoga bas reliefs just for fun. It was a goofy scene.

[**Editorial note:** On divine fiat.]

Zeno at this point has to apologize for an apparent smutty and over handed attempt at sucking up to the libido of the reader. As she has already said, this recent ugly scene was only between humans. Comment: Editors tend not to read manuscripts. That led them to think authors create stories with no plot or characters. Or maybe the cast was "weak". Actually Alice is not the issue here. Her own self is not the issue even if her persona tended to be passive. This is ensemble acting. Our cast is more cinematic than book oriented. And Tends to be strong. Not weak. Also we have do some kind of plot. It is there. It may not be easy to discern since it is complicated but that tends to be the case in this genre. But never mind. Just get "off" on the ambience. Learn to truly dig the scene as story within story, endless action unfolds. Enjoy just Being There. Even if it happens far away. Queen Cal may be a swine but the aliens mean well. Consider this: Zeno and Flavius have **real** power. Cal was under some impression that She Herself was only a figurehead. She did intend to fight for power.

[**Pause.**]

Next morning very late, after food and booze, I woke up in some large room upstairs. It was designed like any modern hotel suite on Earth. This Royal Palace was vast. Covered whole city blocks. Most of it was at least 200 feet tall because most rooms had twenty foot ceilings. Public places such as, say, the Banquet Hall, which was where I first appeared, were covered with domes that rose high over our basic roof area which was flat & paved with gravel. (Yet even this had gardens, lawns and patios open to public.) This Royal Palace of Calpurnia resembled The Pantheon connected to several Hagia Sophias and was surrounded by one square kilometer of gardens.

Through my window could see that foliage here was still tropical, but being on a plateau it was not as lush. Not as hot and humid here as back in Gulf City. Nor was it polluted. Air was clear so I could see mountains in the distance. In fact there was desert south of this area. Due to rain shadow from that tall mountain range. That even had glaciers barely visible in distant haze. I was happy to see snow.

Fascinated by such alien marvels, I stood mesmerized at my big open window which was of local design. Its panels slid into slots in stone walls. They were of tinted glass set in lead as if this were a church. Decided that Fontana Entheos - the entire city - had been built in Byzantine style. Massive walls with icons. I had been told this urban area was only 100,000 years old. Gulf City, the largest was at least 50 million years old but was almost empty, as you know already. Last night had consulted PC built into my vanity desk in my private WC. My tourist guide prog in English told me that this city still has over one million people in it, which have stayed put ever since it was created. So I deduced that Fontana Entheos was the real capital of this planet. Power has shifted from that giant harbor within its ugly urban sprawl to a more livable and much smaller town up here. F. E. was also more attractive. Visually and emotionally. It had far more humans living here openly, as opposed to being hidden. How did I know this?

Simple: my guide in my PR blurb was a cute young girl dressed in the uniform of flight attendant. (White dress shirt, red mini and red boots, etc.) As she stood on that busy street, crowds walked by behind her. She was lecturing into mike. Most were typical Neoforms with their blue skin. But some were Terran in some kind of ancient costume. As this blurb went on, I learned that 20% of the locals were human. This amazed me because I knew that the rest of Xiotan was virtually native and also becoming more and more depopulated. Now less than 500 million were left of many billions. This process had begun only recently - less than one million year ago - but was not reversible. Only a strong human presence had stopped it and we all know that my Race breathes oxygen. Possibly the remnants of some ancient military force still occupied this area, which we shall call Southern Altiplano. Made a mental note to ask Zeno about this. What had been obvious was this: Zeno Galaxa as Empress ruled the entire Eridan System. Her presence supported the regime of Calpurnia who was supreme ruler of this planet. Until just yesterday I had no idea of what any "power" concepts ruled here.

[Pause.]

High above Xiotan we can see its one moon called Zafir. Or Safir? It is a mere 1,000 miles in diameter but orbits 300,000 miles away. Being only 7,500 miles wide Xiotan has weaker gravity. It was otherwise much like our own Moon. See Book Six, Reader's Guide for details. Under its surface were bases which belonged to various factions. In other words, most were loyal to Galaxa's regime, others to Ariana, who only had legal power over some other system, which lately was Epsilon Indan, another orange dwarf in "K" range. Also we are told that most of Safir's occupants are Archaeans. The largest of legal bases belongs to Flavius Inc. Traders who are of Omgal. For hundreds of millions of years Omgal have been building and selling ships & hitech stuff to both Eridan empires. [See Chapter Five.] Fluffo Flavius himself does not live here. He had some meetings on Zafir. It was about those Arian Ops Black who had constantly infiltrated this system. His company - as well as his race - wanted to stop the coming Eco Doom of that wonderful tropical paradise below. So now finally he had joined Zeno on his first visit to that planet, which they also called Eridan Two. Now also they were hoping to meet & train another Terran draftee for the role of messiah to lead their side to victory. It was not to be Calpurnia, but Alice Roanoke.

END OF CHAPTER SIX

Next: Chapter Seven

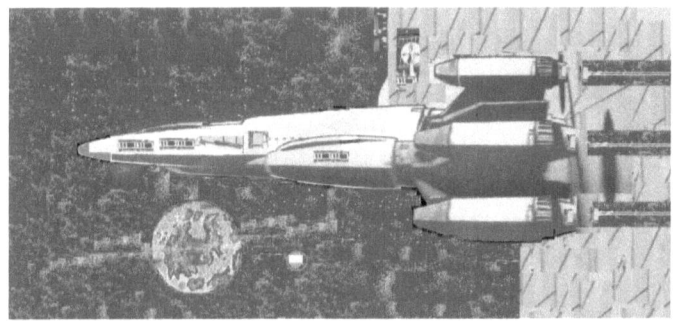

Trader Flavius on the Inter Rosetta which is an FTL medium. Greetings from his people, the Omgal.

Cimora
President of the Planetary Council of Xiotan. Elected as leader of the Democratic Federal Republic of the entire world. Holds office in her Federal Building in Gulf City, their capital. Has only 100 grunts with real guns to save her ass.

"Trader Flavius"

Bam. Bam. Was summoned from my reverie by knocks on my door. Zeno said "Open Up! Its me." I let her in. She was still in her uniform. Along with her came Flavius. I noticed they were alone, expecting that Audrey might be with them. All the better to threaten me again. Both carried Belgian SMGs and Sarin. They were now armed and more confidant. "We have come to confer with you alone, Alice. By the way, that bitch from Texas does not scare me at all, nor my fluffy friend." He waved, said "Morning." with affection. That did much to comfort me. I really started liking both of them. Told them I was glad they came over as "Sorta felt I was at loose ends..." etc. Zeno sat on my bed in her typically slutty posture. She asked me if they could turn my room into a "Consuite" which clues us in on her humor. She has one. So did Flavius, who was wearing of all things a yellow toga. (Oh joy. Life on Xiotan seems to be one big toga party!) Plus his piece. Anticipating my questions he said it was to symbolize Buddhism aka Yellow Magic. "I see all of you have been reading Bonewits". Quoth Zeno, "Forget about Audrey. She has no interest in that. De nada. So leave her the fuck alone!"

"Then why did you hire her?" said I. Flavius, obviously an equal of Zeno's in this power structure, said that Audrey was simply useful to them as tough Terran. "You two have much in common. You're both American, girls and white. But other than that, nothing." No kidding. It actually came down to ecology. Back in 1947 there were no antipsychotic drugs. Like Thorazine. Audrey just had nothing in her blood aside from some booze and pot. She was picked up by Omgalii. Alice had PCP in her. That stuff was detected by Zeno herself in 1978 in Ohio so she decided to draft that Terran for Her War On Halides. "I mean," said Zeno, "usually those with such chemicals in their blood have cause to fight other Archaean cultures". It was finally so clear to me. Flavius like any of his kind breathed oxygen. Back in those days his fellow Omgalii plus Grey Guys, all into oxygen, were the only aliens to visit Terra. Yet, during the 70s slowly the Eridan started to invade our planet then drive out other aliens. Later on Flavius was to explain some of the technical details and was his job but suddenly in a flash of true genius knew this: Aliens were invading our planet seeking any natives they might "relate" to. In those days, from 1975 to 1980 rumors of some cult called Illuminati spread. First in North America, then Europe. Like Aubrey I was no fan of this bullshit. Me paid it no mind. Like real patriots both of us ignored any rumors. But now that I am here, why not consider this: Proper term for this would be "Empire" as in galactic. The Illuminati Conspiracy was a concept for us Terrans who functioned as a sort of link between Terran in general and alien visitors. Period. So now for condensed version of what Flavius was to tell me in my room that day.

[Pause.]

Trader Flavius was Chief Executive Officer of Sirius Industry Inc. which was only one of many large companies run by his race. Called Omgalii, they evolved from polar bears on some obscure planet near galactic center and then developed space travel 200 million years ago. They are not our oldest race by any means!) That planet, called Nascia by them but Omgal by any Eridan, such world twas covered mostly by water ice but had some meager foliage - stunted conifer mixed with tundra - around its Equator. With no seasons that narrow band of vegetation allowed these small beings to thrive.

Circa 130 BC one of their tribes hit our local star Sirius. But that "tribe" was their most powerful Nation. They built a colony there on some dead glode which resembled Mars. They called it Tangerine Dream. That has since been Admin Center of the Sirian Empire, which dominates trade in our small part of Milky Way Galaxy. Such traders used to speak Sumerian, then switched to Greek then Latin. The Flavian family like to speak Latin instead of their native tongue for some reason. They have abducted humans since ancient times to colonize other planets.

Omgal society has no State nor Church. They are run by the Free Enterprise Sytem so that makes them averse to most Eridan who are onto some kind of Socialism. For instance, major stock markets for our Sirian Empire are found in Sirius System itself. They are run naturally by these bears. Here's how Omgal relate to Eridan: Basic data from Exobiology tells us that Archaea are found all over our galaxy. Many Races of the Milky Way are billions of years old.

Some are even of silicone but that is not our topic. Natives of Epsilon Eridani are now spread over our galaxy as well since their Race is 1.5 billion years old. They evolved as Zeno has told us before on Xiotan as beings of carbon flouride. A kind of Freon biome. Again since they cannot have metals nor some other industries due to their corrosive ecology they have had to attain weapons and ships from Omgal traders. To pay for this, they had to fight in any convenient wars. For 80 million years two rival nations of Eridan have been at war. For about ten millennia now, Zeno Galaxa has led her empire in its conflict against Ariana. They are reborn over time. This is where Flavius stops his longwinded lecture.

[**Pause:** Did you hit our snack bar & get a chance to urinate?]

Fluffo was sitting in a stuffed chair while he was talking. He even looks like some stuffed animal. Just cannot get over that! A talking bear is one thing I had never laid eyes on before. Feel like hugging him. That cape makes Him look cute. Like that zoo monkey those Obscurantist Mofos back in 410 Anno Domini had as Emperor of the S.P.Q.R. Zeno sat on my bed bored by this speech. Could make shitty statements, not as hippie but ancienne fascisti. This entire scene was same as always. Warfare. For money. Never Sex. Sex is what we do not sell here in la genre de S.F. C'est la Guerre. 1870? 1914? Fuck you. Were old friends. She looked femmy in her disco diva outfit. Motown anyone? Now I know why Aubrey hates her. One

Trinity of cunts vying for Power over not Men of Terra, not even this Ass Ender planet. Xiotan? What a joke! We are no longer Mensch. So this be an alien Invasion Never did They invade us from any otros galaxy. The Universe is too big! And even so, when Byzantium did dump upon Islam warrior, in 982 A.D. then what of it? At that same time two Caliphates did fight each other. 1978 was rife. Sam the Sham. With phantom armies. Was that so? Maybe some medieval fantasy.

"Anyway" Zeno said "I am sorry for that long bit about how they can support their unemployed even without some kind of public infra structure such as Welfare state. As he said, their Welfare is included as lots of generous pension for work done. But it's only a company fund. Now what?" Alice did figure in her naive Yank mind, so You are God Almighty, whom I love above all else. Woman or Man. Who are these noisy jerks all about me?"

"Fuck you."
"Can I suck up to some radical notion?"
"Yes. 100% if Thou art my amigo. Me Zeno."
"What?"
"Quantum Wave Mechanics. I take and kill."
"What now?"
"Time and Space was mine elder Faith."

Like an ass I asked "What about those who don't work?" Zeno said they did also have support for the sick and fucking helpless, this world is just another carbon copy of ancient Germanic B.S. In other words, their own kind of Marxism. Money came from divisions that existed within their companies as Investment Banks. Euphemism for outfits that simply bailed out idle parts of society. Unlike real banks they never asked for loans to be repaid. They operate in The Red.

"How does that work out?" I asked again.

"They just close down divisions. Not parent firms. That cancels all debts both Public & Private. But since most firms have been going public anyway it makes no diff. Who cares if its state or private capitalism? It works for me. I just like analyzing this stuff."

"Come again?" That was the first time an alien has ever handed me her own personal opinion of anything. Came as a shock. How erudite. Zeno told me that she had to restrain Aubrey in her blood lust. Her main point was simple but basic: We humans had a savage view of violence. We still had some contest between War & Peace. Aliens on the other hand, had concepts that were complex and hard to understand. It went beyond some simple dichotomy of Hawks Versus Doves. They went by subtle gradation.

"Some continuum of gender identity. Neither male nor female?" I joked. Politics and sex? How cool.

"Ha. Ha." spoke Zeno. She smiled at Fluffo. "Us Eridan are all female of gender 100% and we also tend to be Lesbian along with auto erotic. No choice. His kind are both genders and usually hetero. But don't worry about Audrey cause I control her." - "Sexually?"

"Never. Cannot. Her skin would melt from my constant exuding of sulfur dioxide. Corrosion. You cannot touch me. Nor let me breathe upon you. So, even now feel tired because I can utilize oxygen or chlorine but it does not give me the same energy level as flourine. Okay?" She was finis with her spiel.

"Don't worry, Alice." Flavius said "We are here to keep this planet from being altered back into its former ecology. I am into oxygen like you." Then I changed topic. "What about Grey Guys." Zeno told me that they are biots of same origin as any human. Carbon forms who need oxygen. They explored Terra for some unknown race from 1947 to 1977. Then were driven away by the combined forces of us Eridan & Omgal. We call ourselves Galactic Empire but honestly do not control more than a tiny part of Milky Way. Nor can we. That was yet another lie we aliens will hand out on occasion." Zeno nodded. I felt drawn into their empire. Even ventured my random guess that those Greys were hired by Ariana. Both of them shrugged. They did not know. About then, I really noticed that aliens unoften handed out opinions. They had them, sure. In their own heads. From time to time a being like Zeno or Fluffo could mention some opinion on any issue. Both general and personal. Yet it was always done in an offhand way. Judgement was not an emotional matter to them as it is to us. These two never took offense even if insulted. Words did not matter to them. Aliens.

I finally realized that I was chosen for my mind. It had been altered just by drugs alone. At that point felt like asking them about Timothy Leary's famous concept of brain circuits, which I knew of only in a vague manner. During my university days some freaks used to bring it up. Had no idea this stuff existed before 1975. An old theory of mine: Before you analyze something or fully try to understand some New Thing, chances are you will experience it. That really goes for space travel, magic & drugs. For some of us. But what about Aubrey? Suddenly Zeno stared at me & said "She does not count. We needed some healthy person from Earth to work for us. Defend this planet." At that point both leaders stood up in unison. Telepathy? How fucking corny can it get? **This part was pure Con Suite.** As if they were Tweedeldum and Tweedledee. They left my room. Zeno told me to take a "powder" then see them for brunch in an hour. No, we had not eaten yet. They had vital info concerning strategy to get into along with Aubrey. Like be there or be square.

[**Brunch.** Shave & shower later.]

Had my bath while I waited for brunch. Right in the middle of my bath I went into a trance. Then I found myself sitting on polished wooden surface down in some other room. Could tell from its dark ambience that it had to be her Banquet Hall. This is where Queen Calpurnia held court. Sat naked on dead center of their dinner table covered in suds. Zeno must have teleported me again. This time such was even more embarrassing. Seated on cushions around me were Zeno, Fluffo and Audrey along with dozens of servants and armed guards. All were female and dressed in fashions from our ancient world. They had arms from various eras. As in guns from 1850 to 1978. "That was not funny..." I began.

Zeno apologized and claimed that Neophytes like me had to get me used to teleporting. "Then you can do it on your own." she just opined. As I sat there being stunned silent they burst out in laughter. Warm water ran off my body then dripped to floor. Splash. In fact gallons of it had been taken along with me plus lots of soap. "Smells of aloa vera." Audrey said. At least she was now cool. "Do not think you can do this to me" Zeno warned "Also in no way do those present have my powers. You will never hack into any failsafe commands in my mind. Only us real Eridan can fully get into it." I slid over to the edge then stood on floor. They handed me a blanket. As we did this I noticed that our table was clear to avoid having me land on any food with my ass. So I sat down while plates of food were served. Audrey threw me a large towel. Brunch included coffee, tea, juices, wine plus many items like fruit, toast, cheese, rolls, tacos and other casual fare. Even had French fries which, I like. In fact most native cuisine was just a xerox copy of fast food on Earth. This never surprised me. Even Zeno could eat what we did. She was the only Archaean present. Relaxed and enjoyed. "I did not have any nosh since dinner last nite. At least you are good host. But what about meat?"

Zeno said we could order a limited menu of meats. Mostly bacon, sausage or eggs if we wanted any. That came extra. All food was on the house. I noticed she tended to on occasion speak for Audrey. She & Fluffo were having bacon since they liked greasy food. So did I. As we noshed Zeno explained her Race and astronomy some more. For not knowing much astronomy I would not have a chance. Even Audrey was ahead of me. She smiled and nodded while Zeno lectured. Here is a summary:

[**Final cut of Eridan astronomy & why:**]

Her Race had evolved as she had told me back on that mall roof about 1.5 billion years ago - or giga years as they say - on this planet. Xiotan was its native name. Since Io of Jupiter had a similar name and similar Biome, I can see where its name comes from but never mind. Also we knew about that ancient genetic experiment. Now as to how they relate to Terra: From time to time various factions of Eridan came to our planet mainly to tinker with us. In genetics as well. But still not as drastic a change as from an Archaean to Eukaryan Biome. As we know as of 1940 that happened in one gradual process from 3.77 giga years ago when Terra went to an oxygen ecology. Then blue-green algae filled the oceans. Until now when finally we have 20% oxygen with only 1% carbon dioxide. Roughly the same held down here, only since its inception oxygen level has gone down to 10% at this point. It was cut in half by burning fossil fuels and forest. Over in Gulf City smog was easy to notice. But now, back to what was done in Sol System. Galactic politics can be complex. It was only 100 million years ago that any Eridan explored our system. Many of them lived on Io & Titan which were quite comfy for them. They still live there. Usually factions agree. Yet for some time, these two empresses have been fighting over Sol as well as other systems close to Sirius. This was okay by Omgal. Was lucrative.

For now about their names: Humans have called them Xuan, Chuan, and variations of that. The Chohan are now a certain Indian tribe who long ago fought blue "demons" in legend. The Celtic word "Leprechaun" might refer to the skin color of certain aliens. (Such as leprous meaning as if they had a gross skin disease. Blue or purple.) They were also called Fairies, etc. This was widely known. Irish folk tales to Zeno are a joke because no alien looks like that cartoon thing we see on Saint Patrick's Day. What Zeno had to say is what Her Race were up to as our history passed us by. Anyway, the real issue arose 12,000 years ago when some Siberian tribe crossed into New World via Bering Strait. We assume an ice age helped them. Those stone age people spread all over the Americas. Was yawning from boredom when Zeno hit this part. By 8,000 B.C. they smelted copper in Mideast. Later this spread all over. Even to Atlantis.

"What?" I suddenly gasped. They stared at me. Felt like a moron. Zeno said this part was complicated. By 5,000 B.C. they had writing and math in Sumeria. Only 25 letters & 10 numbers. That is where our modern alphabet decimal systems come from. That is what Zeno & Ariana had to learn from us Earth natives just to deal with us. No, it was not the other way around! We did not get our culture from some jerk in a flying saucer. That part was funny. We evolved a very basic alphabet for everyday use. Never mind China with its 3,000 nor Runic with 800 letters. Zeno said cuneiform was called demotic. It was used by common people. Other script like Hieroglyphs such as in Egypt were only for religious purposes. Anyway cuneiform died out eventually to be replaced by Levant, Hebrew, Greek & Latin alphabets. "Also we aliens," Zeno said, had to adapt. "We needed allies." Why ten? Easy. Humans have ten fingers. (Math is abstract 100% as we could have used eleven or even twelve digits in some alien system.) Math is pure science followed by physics on a scale of abstraction. Also Sumeria had decimal system. Yes it was based on standard orbits of our Sol System. Terra & Luna go by twelves or permutations of that and so does Mars. Then we hit Jupiter.

That is when it gets weird. Jupiter has an orbit that goes by fives and sevens. Then we go to our outer gas giants with other Cosmic Rhythms. Such as in that Renaissance ideal of the Music Of The Spheres. Permutations of 12, 5, 7 and 18, found within various calendars of Mankind. Just to cope with all of that, then on to predict outer planets, Mankind had to develop Calculus back in the 1600 to 1700s. That is why ancient calendars vary so much. Holy fuck was that a mouth full. My head swam. Audrey was onto me as just some naif. They were polite. So what about Atlantis? Zeno continued. On ancient maps - using a grid of 360 degrees - Atlantis was and is depicted as two giant island continents. You can consider Greenland also as one of those. Or even Iceland. Romans called our British Isles Hyperborea. Ultima Thule was their name for Greenland. North America was to them North Atlantis. What we call "Atlantic" ocean was to them the East Atlantic with our "Pacific" as the West Atlantic - aka World Ocean.

By 4,000 B.C. some people from our own Mideast were busy exploring Atlantis. They got there by passing by Morocco, down the coast of Africa then went from Senegal to that Eastern tip of Brasil. (It was easy.) By 3,000 B.C. they colonized Mexico. We assume they mingled with natives to become Mayans and Incas. The Mormons claim that in 600 B.C. one tribe of Judeans landed in Yucatan to do the same. They had to be the very last Old World people to do this. For some reason there was no more contact until Columbus. What happened? That entire continent did not sink nor did sea monsters exist.

Plato was wrong. We aliens know that ancient maps with their own grid pattern were lost. That Great Pyramid of Cheops was The exact geographical center of our land masses. We chose Cheops in Giza for marker for our Prime Meridian. Ancient traders used that grid to navigate by. They had to go to Atlantis to run checks for accuracy as time went on. Such data was based on very ancient surveys. Then in 1688 when England was powerful, Newton and his Royal Astronomy Society chose London for their new Prime Meridian. Greenwich Mean. It was 30 degrees to west of Giza. "As you know, Alice it is now your new Zero Longitude." Zeno said. There was lots more which I've left out. Was sitting there still wet from my bath. It had been over an hour now and was getting chilly.

No matter how hot it gets outside thick stone walls can insulate you well. Thus, most ancient houses needed no air conditioning. They also had big windows which tended to be left open. Could feel a draft. So

I interrupted this lecture to ask to be dried. Audrey clapped hands. A servant came in with some beach towels plus a bowl of lemon water. They poured this water over me to rinse me off. Then they dried me off using towels. What "royal" treatment. Then I finally sussed that Audrey was okay by me. It was intuitive. Happy to have short hair which was just easier to take care of. Zeno was busy opining on the alien tendency to Be evasive. Like, after all this Hollywood stuff? Whish Types like me had been exposed to? Well, as an answer, which I and no doubt Calpurnia must agree with; OK: Had They wanted to invade or destroy us, They could have easily done so, openly, long ago. No kidding.

Zeno took the floor again while all of us listened with eager alertness. Had finished eating but was still into my third cuppa coffee, which was triple "A" quality. Cuisine in this palace was tres bon. Issue One now was what was wrong with natives of North and South Atlantis - the New World. It was clear Calpurnia was in control as Master of House. Yet she was just sitting there enjoying her meal and Zeno's lecture. (It was likely that only North America was called Atlantis and South America was "Mu" or even "Lemuria" but that was no big deal.) These tribes for 12,000 years had tools of stone, bone or some base metals, but never had any kind of steel industry. (Which needed blast furnaces, higher step in technology than mere copper smelting.) They also never used gunpowder. All of this is known.

Only why? Well, she had an answer. It was Ariana's agents from both biomes - who had paid frequent visits to Terra. All of this time dey experiment in DNA and culture. Any humans within Old World were allowed to develop de steel y gunpowder as needed, while such were totally forbidden in all parts of our New World. Even Ultima Thule). As we know, iron was invented in 1,300 B.C. by some nation of the Mideast. Then spread all over. Then in 800 A.D. gun powder was invented in China. So our invisible cultural barrier was across Bering Strait. One more point: Their New Date Line. Now at 150 degrees West. Through Anchorage, Alaska, then onward, East of Hawaii, then South, far away, thru Tahiti. Now runs thru Fiji, which is at 180 degrees. Recall how our Prime Meridian has shifted after two millenia by 30 degrees to the West?

"So Ariana was involved in keeping Inca, Maya & Azteca tribes down in her own way?" I asked, "and how were such things done?"

"We are sure of this. So were our Sirian allies. In fact, followed it since inception. We also know they used Biots to visit Atlantis and so forth, meaning Grey Guys. We assume whenever some native city was close to creating iron or gunpowder they bombed these poor souls back into Copper Age. Then they even tried to return just to find out what had happened - in 1947. But we got rid of Them by 1978 with help from your USAF. Terran warplanes shot down Grey Suocoupe Volante! Easy since their level is only 500 years ahead. Of Earth in 1947. Okay? That is why Ariana hates us. We fucked Her nerdy experiment up Her Royal 61 Cygni Ass!"

"Why did the Vikings fail?" Zeno explained they were too back ward. Simple bandits with no charts. Pagans. Columbus was smart & powerful. He had support from The Church, Spanish money & ancient charts of that older grid pattern he was given by refugees from Byzantium, who had fled to Italy after that Turkish invasion. But you know this stuff already." Said that her story explained much of Earth's politics as of 1492. It was due to alien invasion.

"Right. More or less. Some of us even left some lousy effects behind us by accident. Some of my agents wandered about in your Equatorial regions in 1975 & as they did so, they urinated on plants there. These viral microbes - we are all based on our RNA being Archaean - were deadly but only to natives of Terra itself. Like H.G. Wells in reverse. Some of your worst virus epidemics have evolved from our own body fluids. Gunk we aliens left behind. Even Fluffo here whose DNA is closer to yours - much closer - is a danger to you. Your immune system. So you see, we had to vaccinate you while asleep plus put sulfa drugs into your food. That went on for months just to allow you to live here. Same story for Audrey. That ends my lecture."

[**Pause.** Wow! That was really pyrotechnic. Neat how the coffee always tastes so good on other planets. They must grow it right there. On those misty mountain slopes close by Gulf City. Well?]

My note as Alice: Brunch was over at 13:00 Hours Local Time. This time we walked to our rooms. Fluffo was somewhere on my floor. I was told to relax and wait until 17:00 Hours, then return to our Banquet Room. There was to be a dinner for all plus more speeches by our leaders. Audrey had grown nicer but had little to say. So I rested on my bed to work on my notes. After some time I fell asleep. My anxieties had vanished. And the AAA+ coffee helped me live though my ordeal. Must have!

Had some funny sorta dream. Deep in REM sleep - had reported on sleep experiments a few times - was sitting in the waiting room of Psychiatry Wing of some hospital over in Tel Aviv. In my usual war correspondent outfit. Combat shirt, over blue denim shorts, tee shirt noire and sneakers. Also had press pass pinned to my left lapel. Reading articles on schizophrenia from a stack of Psychology Today magazines. Seemed a new fad. P.T. mag was full of stuff on "shizo" this and "shizo" that. Along with Teen Warriors, Deadly Street Thugs, Our Growing Menace of Gang Related Behavior, Multidrug Abusers etc. We seemed to be in danger "again" from Within. Aside from evil Arabs or Russians.

Boring, yes. This National Bore. All around me gimpy types wandered on foot or in wheelchairs. We were all in some trance from pervasive Anti Psychotic Dope. In massive doses. Deduced that it was always put in our food. Even me. I was here just for minor injuries. A distant female voice called to me from above. It said "I am an alien. We have detected halide gas molecules in your blood. You have been selected." For what? Or so I wondered as I sat there. Then woke up. Back in that hospital. It was valid memory. Back in Israel.

Empress Zeno Galaxa loomed over me. She must have been standing by my bed for some time. (How did she get in?) She told me it was close to dinner time. Invited me to come along with her just to have a short and sweet discussion before joining los otros. I was already dressed in shorts and halter top but barefoot so that was okay. I had long since gotten used to alien creepy behavior. Besides I'd wanted to find out what Zeno had to say. T'was usually more meaningful than what most of these beings had to say.

We ascended some dark, narrow, circular staircase of stone. Sand stone. As if that was a vital fact. We came out in a large chamber high above our Banquet Room. Above me our roof curved as that huge concrete dome which formed our tallest. Floor was of wood supported by steel beams and very old. There in the center was a chess game of 18 by 18 feet. This was divided into exactly 64 squares. How logical. Just like any Terran chess set. On it were pieces of lacquered wood, brown versus ivory, each three feet tall. Pawns were carved to look like Omgalii. Rooks, bishops and knights were regular, but queen and king were also replica polar bear runts with cute capes & crowns, etc. Broke down in helpless mirth. "I hope the fuck Mister Flavius does not know about this bullshit!" Zeno smiled. She had taken off her booties - called 'em Fairy Boots to her face - but still wore her usual cosmic bikini. She was also barefoot just to be comfy. "Check out this floor" she said. I walked on top of this set. Made of thick glass with alternate tiles in clear and tinted glass. Latter tiles were a smoky grey.

It was clear that we were to see down onto the floor of Cal's Royal Banquet Chamber far below. About 200 feet down. Peering through those tiles we could see some familiar figures sitting around our main dinner table. A feast was already in progress as it was by now 17:00. Cute. My respect for alien society had risen in five minutes. Zeno said "No doubt you've heard of 'Dune' by Frank Herbert... so in that series they have Harkonnen - the Badguy - make his prisoners stand upon some giant chess set like this one. Only this creep has each tile be a trap door which falls open each time some chess 'piece" is lost as a move in their game. Beneath them of course lies a watery pit full of some kind of moat monsters so that our poor victims die by chance. How evil. Eh? That is my own lecture."

"Really? But you would never do that. I think."

"Never" Zeno replied. "We are into real activism here. Not some brainless decadence. Ergo my point." She then said that this chamber with its chess set had existed as a minor luxury for centuries. Earth culture had for long been an influence here in Fontana Entheos but not elsewhere. She then led me to French window which she pushed open. Outside was clear blue sky. Fresh breeze played over my face. That was refreshing after this stuffy room. On a small balcony bolted to brass rail, was a coin spy glass provided for tourism. Zeno invited me to use it.

Over vast distances we saw tall mountains surround this plain. The highest on our planet was called

Emperor Mountain. (Peaking at 25,000 feet it was covered in snow.) We walked around this dome on a narrow parapet to come to another telescope on our north side. This 'scope revealed endless plains flat as pan covered with selva. Far off in haze lay Gulf City. My home for months now. Over East was more forest. Down in swamps, right on Equator lay the ruined city of Tapiran which had been their original capital. Pyramids hidden by foliage. Giant hulk of stone over 80 million years old. Some huge pyramid, far taller than our own Cheops, had once been marker for crossing point of their Prime Meridian with the Equator. Tapiran was empty and haunted.

"What are those?" I suddenly asked, having noticed dark clouds of smoke rising from many spots all over that green plain. "Fires" Zeno said, "Our enemies are busy destroying forest as much as they can. They are trying to increase our Greenhouse Effect. Just as I have already told you. Now see for yourself." She pointed down to gardens nearby. "Look." I saw some choppers parked on some parking lot below. There were familiar Chinooks but also VTOLS. All painted in baby blue. "They belong to our Planetary Council." said Zeno. As you must know, our own are black. Ariana has planes painted camo green. Okay?" We now assumed it was Cimora paying us a visit. Zeno must have expected her. So here was yet another new person to meet. Some V.I.P. dropping in for dinner. This interlude up here was meant as briefing. Then at least I was prepared. Zeno told me to follow her down below. For dinner. "Do as I say. I can handle this kind of thing myself. Just relax and you will be okay." Life seemed to be some new routine but was it comedy or tragedy?

[Pause.]

Zeno and I arrived in our Banquet Chamber at 18:30. That was 1.5 hours too late but they did not mind if we were "facionabel" late. Flavius was there along with our host. Those two had lots to talk about anyway. Their servants kept them company. Here is what had happened just before we left that room up in the very top of our dome: It had become dark so Zeno still wanted to show me more wonders with those scopes. Now both of us were turning into real astronomy buffs. First we looked at Zafir close up. It shone with a very high albedo. (Like our own Luna.) Up there in some illegal bases our common enemy lurked. Then we checked out Eridan One that planet closest to this star. That was much like our own Mercury, a dead globe the size of Mars. Its local name was Nebodan. We also had look at the two gas giants plus a wide glowing halo of yellow freon gas out beyond 30 AU. Zeno told me her kind had to live at high pressure to stand a Biome of 250 Celsius. The original, aka long before Day "X" atmosphere of Xiotan had been ten times thicker than Earth, consisting of halides diluted with nitrogen, argon, lots of carbon dioxide and even sulfur dioxide. That was of course toxic to my kind. Then we walked back down to dinner.

We approached our main table then sat down on it. "Hey jackass" said Aubrey. "What kind of manners are those? You parked yor butt next to food." Indeed it was so. Huge plates of hot food and drink were laid on. There was meat, starch dishes such as rice or potatoes plus salads, bread & fruit. Decanters and some urns contained many drinks. There was even beer and wine. I got up, declined my comment and asked to be served. Our waiters in skimpy gear handed over silver platters of my fave items. (It was fried chicken with mashed potato and cole slaw.) As I ate, all others continued their talk. In her democratic way Aubrey allowed servants to eat along with her. Some of Fluffos clan were also there babbling in Latin mixed with bad English. There was even a band playing strange kind of jungle music. This was fun. Yet while I was busy stuffing my face I still noticed that Zeno was out of sight. Was she hiding? Soon a noise came from our main passage leading to front entrance. In dim light were seen a crowd of some kind walking towards us. With typical lack of reverence, I felt like saying "Welcome to the Real America: Land of true comfort food!" Then dropped my food. Panic grabbed my gut. Chicken?

More than fifty people of some kind entered. It was hard to tell what any of them were due to glare coming from behind them. (So much for alien methods of heating, cooling and lighting.) They were loud and armed. "It's Cimora and Company." said Flavius. He sat at head of table just cool as cucumber so he could comment on The Action, as it were. (Was he into sports? Hmm?) This looked like a Triple Header, whatever that was. (I am not into football.) To be serious however, what we saw was a crowd resembling ourselves. Most were Neoform natives, others were real Humans from Earth, with a few renegade Omgalii. Motley crew, as they say. They were all in the same baby blue uniform of our

Planetary Council. These fifty odd mutts were a SWAT Team assigned to Cimora for her own personal protection. (They were the only group on this entire planet licensed to carry firearms to keep Law & Order. Gulf City Police - as we know - simply had nothing but stun guns and mace. Wow.)

I assume they were the only military this regime had. No wonder Cimora must be pissed off. So then we just played the Innocent. This seemed to be my Ride Over Lake Constance, an Earth opera. The trick is this: Any real European of means should know that that lake has its own tropical micro clime so how can you ride over it? Even in winter there is no ice. We all felt real nasty conflict just brewing. Between Cimora and Calpurnia. With Zeno involved as a resented broker. To the death we should think. I sat next to Flavius. "Do you ever read Valerian? I did say just to ease tension. "Like, do you ever get tired of Laureline bossing her Beau about?" Flavius quietly agreed. He nodded. "There is some threat of mayhem. Okay?" Now I was right on to be with the real power behind what they call The Sirian Empire which must be erroneous. It was not even a cluster but a region. "You will get technical with me?" Cimora yelled out. She was leader of this armed phalanx. Behind her they stood at the Ready. Rifles pointed down to floor or up High Port. It was a staring contest. Not yet lethal.

"Jerk me around will you?" yelled Cimora in deep shadow standing in light. Like God. "Am President of this Supreme Planetary Council! I am not a... never mind. We shall not tolerate this moron called Calpurnia aka Texas Carnegie one more second. We are a Democracy and have been for 80 million years... no Royalty nor otros Aristo here! Capiche?" Cimora stood 175 centimeter tall, was of slim figure but still was an Eleven as in very wide hips, wide shoulders and pointy tits. Wore simple mini dress of white cotton with heavy plastic belt around her narrow waist. This held hand gun in white plastic holster. In one hand she held High Port an M15 Carbine. Her face resembled that of Jane Fonda, which was typical of most locals. Her hair was short and black. Also in fashion, had lots of expensive trinkets gold y silver. Pearls. Shoulder epaulets revealed Her Rank." Her legs had on white plastic waders similar to what Zenos wore. Again in fashion. Yelled "Me fucking taking over! Where is my goddamn rival? Cal. Show yourself!" At this point not one soul moved. Asked Cimora if her rifle was loaded. She nodded like an idiot. Stalemate. I looked over to Flavius.

Flavius stood up; walked to Cimora. Said, "Me only Envoy kinda thing. Like in that yo TV show. A Thing with funny ears. You know? Like onto Manchild, as mammalian boob thang, Me create some Alliance. Me. Y Zarcon ca esto me fra. Me Caesare Superiore! Sirius. He doth rule! As Known Space. No mean feat. We achieve Greatness! Warrior of Mankind! As Neil Armstrong, in your name. You are one of us! Now have transcended cowboy y ranchero mujere mystico. Amigo. You good!"

[**Alice:** You know, even Flavius can sound like an idiot On occasion. Is he sucking up to Zeno? Ugh.]

Invader, maybe even profiteer? That show you love so much stinks!" At which point it seemed to me, diplomacy, even with Flavius doing it, may not work. It only takes one bullet to make a point. Was this the first shot of some new war? What is wrong with these people? Asked if her name was "Seamoria" such as our English version or maybe "Cimora" which we all took to be Latin. That was just to ease tension. **I was making it up.** Cimora yelled "Coward! Bitch! Were it not for my conflicting biochemistry, you cosmic deviants all woulda just gone to fuck each other. Decadent! Me be legally elected Leader of this planet. Am not from Terra nor some yellow Gee Two Class Star! Bla, bla. Yes, I am the President. Your one and only! President! Who wishes to argue?" At that tense action point I said "How do you know?"

"How do I fucking know? Really! That was some verbal trick of intimidation. Grew up here. No more of your Terran ways. No Third Circuit. Reeification. Intellectualism. Nerdology. Merde! I take over. L'etat est moi." At that moot point when nobody moved a muscle, Flavius stood up. He calmly orated that some nasty conflict which he did not foresee, as he had no ESP had to erupt between two Rulers of this one planet. That was between Calpurnia and her rival Cimora only. Flavius and I were not aware of any local conflict. He meant it. So I agreed. Cimora stopped grimacing in blind hate. Her face relaxed. She lowered her piece, then looked around. "I can negotiate. Men, get back to your choppers. I shall stay alone to parlay." That made me happy. Both me & Flavius, whom I really started to love in some puppy like way, resumed eating. Our kind of accomplices as slaves also continued their meal. Why not? It was far past midnight by now. Here was one factoid in our minds - all of them - but not in Cimora's mind. She was not "in" on parts of Zeno Galaxa's nasty character. Zeno was well endowed with many occult arts of which even Cimora knew nothing. For you see, my Empress had been hiding in darkness behind some pillar. She snuck on her toes around Cimora in wide circle until she did reach a point right behind this President.

At this point you may wonder why Cimora's people failed to see anyone of us sneaking up on their leader, but Zeno was hiding in dark corners all of the time. Only I noticed anything anyway. There was lots of stuff to hide behind. Keep in mind that all of them were still blinded by sunlight. Nor had they ever seen Galaxa in person. (They must have been alerted by agents of Ariana. By the way I had often suspected Ariana of having fomented those bandits back in that mall. Maybe they had recently sent a report on all of us directly to Cimora in her Executive Palace.) Aubrey was her rival. Zeno an agitator. That made all of us one faction.

While my friend was hiding behind some partition, Cimora ordered her team to return to their choppers outside. "I can take care of these losers" she said. None of us had any guns in the open. Then she just asked where Aubrey was. I looked around. "She must be taking a piss." Some servant said. Cimora sneered. How rude. Suddenly, to avoid making noise Zeno floated up into midair. She was talented in levitation. Silently she drifted over towards Cimora, who was still standing there aiming her gun like some robot. "I am waiting. Go tell that bitch Calpurnia to come out of hiding ASAP. I may just waste her on sight. She cannot escape. My men can be looking all over this dump. It is sealed off totally." She was cold and hard. While in motion Zeno had picked up a brass amphora. As soon as she was directly over Cimora's head she laced her with this heavy object.

[Dramatic pause.]

"I only knocked her out. Bitch ain't dead yet." qouth Zeno. With smelly sulfuric acid pouring out of her skin as sweat, she stood frozen in posture with this silly looking brass object d' art in her femmy hands. Let it drop with some summary "clung". Then danced in victory. Alien tempora and mores. All I know is that, as both me & my buddy Flavius knew, we can always trust our own Empress. She was gifted. These mundane locals were not. That was just Her complex Path. The Way. Tat twam asi. Our Empress was gonna explain it to us later. Zeno joined us at dinner promptly to have Herself vino in stem ware. As she drank deeply of this, for Archaea enjoy alcohol, I saw how fragile glass melted in her satanic sulfurous lips which I did not on pain of death ever touch. It would melt my face. She drops molten wine glass on our basalt floor. Crash. Tinkle. It was over.

Calmly Empress Galaxa announces: "Was only some palace revolt. Cimora will be flown back to her office in that Federal Building over in Gulf City. Revived, install again like some PC prog. She is a local leader. I need her." Then she picked up her corny Motown silver booties, pulled 'em on and walked away. "Flavius will you explain it all. Am losing my lingo. Daisy, daisy, tell me?" In full kinda Imperial cosmic uniform... duh...?" then she walked out on us. Well, we all know her name. Bequeathed in some greasy paen to pyramids upon Nile? Will never know. **Note:** Of course Zeno also confiscated Cimora's gun. That was her first act after her coup. Besides joining their party. She enjoyed their cuisine. Yet later on that night, she took over Cimora's chopper squad, plus her own and returned with her prisoner to Gulf City.

[Real strange interlude.]

Some kind of Consuite Affair. Mundane. But within molecules of Space & Time? Of dream or waking reality? Suffice it to say in terms of good Art - a Romance - that our Good Guys won? Cimora was unmasked as we may say, being secretly in pay of Ariana who was still trying to destroy as some deliberate Act Of War this ecology of Xiotan? Why ruin established biomes? Zeno Galaxa had to spy by ESP to find out what was wrong with the proverbial weak leaders of Xiotan. So now my story ends. We will only give you another brief speech by Zeno to just clarify relations between us Earthians and Aliens. Only: Do aliens live in consuites?

[Pause.]

Well, I am Alice Roanoke who is an American who is not some kinda philosopher so whilst I believe in God Allmighty, Nietzsche did not in any way deny the existence of God but merely meant to say that some yuppies who were done drunk on power and money, as industry grew during the early 1800s gradually lost interest in their religion. That meant every faith found in Europe, not only the Chosen People. After he died, his sister took his books and promoted them as being Anti Semitic, which created a sensation. She made money that way. That was social Darwinism at work. We can easily see how that led to certain results fifty years later. More or less, in its original form, that slogan meant, actually, that God is "dead" in the minds of the people.

One day, told Zeno and Flavius that I was tired of them barging in with endless lectures on alien philosophy and their **slanted** version of Galactic politics. So they even had Quantum computers and could leap from star to star. Just in my basic way, I needed to know what they wanted me to do, when or if, we finally got to return to Earth according to their Cosmic Mystery Plan. You see, I finally cut short that speech above. My Empress did clam up. She apologized and did leave my room. Later on that day, which was soon after that arrest of Cimora, Flavius invited himself into my room. He started his visit by telling me what had become of Her. She was back in her post of President, even with her usual powers, all Of them being Secular, safe in the "bossom" of her Council. [?] Back in Gulf City with some legal papers to obey. Well, Restraining Orders. No big deal. Well, they be all happy with us. Fluffy was easier to deal with than Zeno. So what about my mission? I asked. To travel with relativist velocity over time to land in a year called 2084. Yet more of such things later. First some small speech by Flavius before he fucked off in some haze - or would UFOs of his own people pick him up? This is what he said:

"Foggy mental breakdown and why don't we come down to it: alcohol and drugs in lieu of sex. Real affection. Love? Anything. Just another mammal who's into sex & love one way or another as you Terrans are. As for Zeno, she is your ally but in no way resembles you. So just let me tell you my story, so you can judge all of this. OK: We deal in arms and ships. This has to be millions of years old. We keep these aggressive jerks – The Eridan – going on forever with their warfare. It will happen again. Reincarnation. You may even know this. I do! More than you! Well, okay, so we got more database. We have quantum computers! Data was stored on crystals going back forever." So the Omgal said. Her teddy bear. Wow.

These beings must be on some higher energy level. What is their concept of War & Peace? This may sound corked, but we cannot understand them. One side wants to preserve us Eukarya, the other not. Who do we then choose? Anyway listen to my story. Flavius had an interview with Alice one day while in her room in The Grand Palace. It was about her three run ins with Officer Norom back in both Chicago and Cleveland. "Where was your first, like Encounter of the ha, ha, Third Kind? I mean with Norom?" Relaxing in her bed, Alice said "On that dance floor in AON Tower in June. At my Con. Next, my home town, in some bank. Later, on my Xmas holiday in Chicago, same bank. But in the AON again. So what?"

"Think about it. Your AON Tower must be very attractive to aliens like their Empress. Why? Cause it is made of white marble and very tall. Of course, only its jacket. Yet it did resemble Cheops in Giza as in being clad in thin slabs of white marble. Did you know that?"

"I am not into archeology. Again, so what?" Types like Fluffo can be irritant. Flavius explained it as subliminal. Aliens are not attracted by our words, but Units of Meaning. Even so, if they perceive our words, they still fuck up any meaning involved. Like, see it this way: Zeno groks on her monitor some event going on down below last summer. 1978 was big for scifi but also disco. Fads. Fun. It was up Her ally. Royally as the saying goes. Even played, by coincidence, that goofy song about King Tut from the new Steve Martin comedy album. (Not on CD yet.) The entire setup as symbolism reminded Zeno of her core persona. As it had existed over millennia. Sumerian and later influences. That amusing concept prompted Her to send one of Her officers down to accost Alice somehow. Along with others. Phantoms in UFO events are often offered to us Earth natives instead of solid ships and so forth. Easier to generate. Alice at that point was elated. "That rings bells like crazy!" She thanked Flavius for his time. Added data to her files. By some accident of perception that ancient complex of Giza attracted attention. Before it was abandoned in 400 AD it was still functioning as huge working city which was maintained by crews of staff. Land around it was not just barren but had gardens irrigated by canals. That included large plantations of date palms. This area of dark green contrasted with Cheops, being clad in those days with white marble. So even from out in space it stood out for aliens to see. By some coincidence Zeno's people noticed one white marble tower among many in Chicago in modern times for their action to commune with us natives. That was our AON Tower which was over one hundred floors tall.

So they used their own advanced concepts of Time & Space to somehow "speak" to us back then. What we see over there in Egypt near Cairo is just some random mass of sandstone. Its marble jacket has been removed by vandals and now lies surrounded by sand dunes. The same dull color. Flavius was a humble being. A small male bear. He told me that larger servant of Zeno's was just another of her many citizens. It was able to teleport. According to data from some device they had which we may call **quantum** computers. Stuff we can't possibly make for centuries! Too far ahead. That naive soldier had noticed very simple coincidences of design in Bauhaus, Mies Van der Rohe, Cheops, etc. Form after Function. Dark green against white squares. And so forth. Subliminal. But it responded. Same thing with how We aliens commune with you.

Alice. Aubrey. Sears Roebuck. Roanoke. Or maybe... give us fone number or legend of Ruth. Howl. Ginsberg. It kind of works out. This was a random pattern. Bottom line was that Galaxa our Leader. Wanted. Some fresh blood. To lead her own assault on Epsilon Indus. That was some third star system we knew was involved. Yet another In her ancient warfare. Then if all goes well, we shall promote you as Empress Roanoke. But for now we stay. Dinner will be over when Zeno comes back in tomorrow after New Dawn to take you away. Comma. You Alice goto Terra. Yes! Flavius my wee bear walked away. This party waxed and waned. Darkness closed. Fini.

[Pause.]

After President Cimora was placed under arrest by Galaxa her forces landed in many parts of Fontana Entheos. That big city owned by Cal. According to myth, say, where some Fountain of Youth was located. In choppers and trucks thousands of our soldiers came from safe camps hidden out in forest. They occupied this city. They were welcome to local civilians as liberators. This part may sound corny. You can tell by its lack of "hip" talk. Well, this was dictated to me by Zeno herself. After our somewhat ruined dinner party we were told to go have some sleep.

Next day Zeno came into my room with this weird head fone device. It was attached to some klunky data storage machine utilizing what she called a "holocube" which was just one carbon crystal one inch to each side. This fitted into a slot. It worked as quantum computer. Her term. No human has ever seen such things. Directly into my brain it beamed lots of data on astronomy. "Stuff Mankind will never know" she said, "Which you will keep to yourself. It will help you navigate vast voids of Space between stars, velocity plus orbits within systems. That will allow you to teleport on your own safely to some point on Earth. Later on you will impress some Korn experts at NASA in Houston with your advanced fund of memorized data on this part of our galaxy. It will fuck up your head a bit but that is part of our bargain. [**Asshole!**]

You are in with my regime and we happen to be winners." Flavius came in to explain their motives. It was simple even as an epic tale. A long time ago, some other star system had been property of Ariana's nation. It was the yellow dwarf Epsilon Indus, very close to us. But in a long period of war fare, our side had finally managed to destroy all of its life. All of it. Down to microbes. Its main planet "Nerus" was now devoid of life just like our own Mars. Genocide. "This does not concern your Earth people" she said. Flavius nodded. "In fact your nation is our ally. We wanted you to join that first NASA mission to another star system which must happen about a century from now. That is my plan."

Flavius said "She has my blessing as it were. It may even be proverbial good story." Sounds like some goofy fairy tale to me, Felt like saying. Them corn balls had de nada for humor. It stank. I even felt like making some stupid comments about our last war over in you know where. That "thing" with Nixon & Watergate. My idea of Journalism my work, is to be objective. It suddenly came to me that all Me hadda do is dig notion being Space Reporter. Began to add up. Why not? They wanted me to fly off to some alien system and then... Just Conquer! These two freaks must think I have a mind like a ten year old. Just some kid. I even knew how our victory was to be achieved. By teleporting into the ships of some "inferior" race. Like home invasion. Then fly stolen ships to conquer all over. Why not? I was never to return to Xiotan again.

In May of 1978 Alice teleported herself back to Earth. She landed safely in the European city of Essen. There she was contacted in a safe house by her friend Galaxa. Read about this in Book Two.

[**Note:** Flavius and his Race the Omgal are total fiction. So are Lizards from Planet Cetiwana, which appeared in an earlier book. Grey Guys and our various types of Eridan may actually exist. This series is a work of fiction.]

END OF CHAPTER SEVEN

Next: Chapter Eight

"Return to Terra"

Here I am again, Alice Roanoke. I am still only 26 and the date is now May First of 2084. The aliens call it Day 130 of Year 21,394 of the Eon... and Zeno just interrupted me to say "Sidereal! Sidereal Eon, Alice." We were sitting in some vacant and very old dump of a place in some part of what seemed Europe to me. It was one dozen floors tall and had no roof, so we were on Stockwerk Zehn, the Tenth Floor. It was some brick office block at the gates of some ruined factory, say auto assembly. Zeno told casually that it was built in 1720 which for Europe is nothing. "I thought they made cars here? I mean like, in the 1700s they made autos? Get real."

"They first made some other shit like steam engines, or what not. Then they renovated and retooled. Okay? How do you feel?"

"Totally fucked up. Acid, man." I replied.

"If we were men, this just would not happen. Us dames can be easier on each other than them dudes. How do you like my US macho talk?" Let her know polite that I was aware of her joking to make me feel better. Looked around, and deduced that this room and probably the rest of our factory compound had been abandoned for decades. Focused on distance to see that our compound was at least one square klick across. There were large brick and concrete hangars that stood over concrete expanses covered in foliage, result of climate change. The walls were overgrown with lush flora all over. There were gigantic trees such as oaks and ginkos. Before all else, this was

Terra. No doubt at all. Could even see the name of "our" firm painted on ruddy brick walls in big white logos: Chrysler Firma Autowerke. There were rusty stacks of steel and glass scattered about the yards as well. This entire compound was surrounded by brick walls topped with barb wire. Zeno grunted and warned me with an unkind burst of ESP. "Do not ever blab out any possible jokes about death camps. I'm in no mood to be senile." That gave me real migraine agony. Like a hot band of steel pressing on my skull.

"I am a serious person," I blasted back, "but Zeno dear, thank you for your sensitive nature. I need your concern right now." Zeno told me verbally to lose my nervous tension. Which she sussed in me. Got up off the floor then spread legs apart, giving me a solid stance. Then shook all over and vomited right there without qualms and finally after few minutes of this it stopped. Still shaking and in a cold sweat, ran to a nearby washroom and relieved myself from every orifice. Then I washed my face and hands and returned to our common room, still in my combat uniform. No underwear, just my familiar baggy gear. It had always been olive green. No flashy camo stuff. It was soaked with salty sweat. **Reaction to being teleported in from Space. Light years in stages. Again, within a week.**

Then I noticed that the climate here was as hot as back on Xiotan. Then I recalled being back in Suite 6610 just before our last **teleport ation process** had been activated. It had been a major effort, jumping in stages from one star to another across one dozen light years, finally taking days locked up in some small capsule, gearing down to the surface of this planet. I then looked around this room to see why it had been chosen for Touchdown. It was 60 by 100 feet across, done in typical 1940s non descript decor. There were various desks of wood and metal in rows. Shelves of wood and glass lined up by walls and windows of thin glass. Most of the small panes covered with thick grey dust while some had been broken... for how long? Had no idea. Posters, calendars and other things covered walls. Anything from auto ads to beer ads to travel ads. House plants in ceramic pots stood on sills which was so European and homey. I had been to Europe and the Levant myself in my old life during the 70s, as you dear readers know. Made myself hot coffee from supplies found here. That helped me get back in shape. We explored our new home.

More detail. The house plants were still alive... so who cared for them? It was strange to have as my next idea. So I assume some entity lived here. This still did not look fit to live in. Chairs scattered all over, some lying sideways. Then cups and bottles which had been sitting there for god knows how long. Desks also covered in thick dust. It was very fine and gray. Zenobia herself stood in the shadows like some vampire, now lurking in full Empireal uniform. It was still her titanium bikini. Her deep blue face was thus obscured while her eyes lit up brilliant orange like a cat. She had that familiar evil smirk on her cute lips. I knew what was coming. She moved mouth and intoned "Well? What is the first thing we should notice?"

Since she could read mine me said nada. Her posture had not changed nor her rude smirk. She had been standing for 20 minutes in one place with legs spread wide and arms akimbo like usual. Asserting her Position in Life no doubt. How typico. Then I saw that some of their machinery was from my era, such as normal ones which were electronic. Those pastel Smith Corona monsters with steel balls. (Ha, ha.) They still weighed a ton. Then some manual ones as well. Then finally, things with separate keypads, CD stacks and finally with a shock, and this tells you something about me, flat monitors. I actually touched them in awe. No power. Zeno stood in the shade while raw sunlight poured in. The glare from outside was intense. I had always liked the shade myself.

Zeno told me to sit down. "Check it out." she said. I did so and saw the words Siemens Aktien Gemeinde on it. But the screen was flat like nothing I had ever seen before. "It's just an LCD screen, Alice," Zeno said. She came over and put her hand on my shoulder. Her smile was gone and she was back to her usual dead pan facial aspect. It showed concern. The third one was the worst until I had finally figured it out. You see, deep frown and penetrating, almost hateful stare was her third possible expression. Facially we mean. I have told you before about the gestures of her race, and with them, body language is the most common way of communing with us humans. That means of course deep and intense abuse of ESP to, well - beyond the point of decadence. And so we modern Terrans have some idea finally of why these creatures are so prone to warfare and total collapse.

They form a culture giga years old to be sure but still, why bother? It was all too complicated. Feelings for my Queen Of Many Worlds overflowed my mind. Went catatonic. Please don't call me a Lesbian. It was the perils of hero worship. Who knows what lurks Out There?

Finally she stated in plain English verbally that we two were really on Terra in the Year 2084 AD and that as I could see, the future had long since passed for Mankind. These flat screen PCs had become common by the year 1995 but not before then. Had been invented by the 1980s, but were still too pricey for even corporate use until "...as we have just said. But we are not here for nerdy talk on PC advances in technology in the First World, or whatever you reporters usually talk about! This is just your previous life which as you can see, did not get further than the Fall of 1997."

"Why did you lie to me?" I asked.

"There was no lying involved and besides, as the leader of my nation I have the right to tell lies, plus I tend to lie personally anyway, and Alice, you're a female Terran so lying is okay anyhow. I control you in any event."

"When we teleported from Xiotan last week I was under the very strong impression that we'd arrive in Ohio and in July of 1979. Oh and by the way, even for me as some young hippie, of all fucking kinda people, the Seventies were... The Future! Even in high school my pals all said, This Is The Fucking Future!"

"I cannot recall nor do I care." the Queen of All Many Worlds intoned in her usually solemn and polite manner like some hypocrite. A funeral director right from 5000 BC in her whore outfit. That was my private schtick. I am me and Zeno was still some alien jerkess I just never understood. Amen.

[Pause.]

I had to slow down. The fact that I had time traveled which was weird enuff for even me, was not okay because me somehow knew such cosa be not "reversible" for some casual reason. I half expected Zeno to comment on my spelling, as in how a word like "enough" was by now written and spoken as "enuff" by my Middle American mind; that sort of thing. But no. She refrained. Heat got to me in a rush. I sweated more salty water into baggy gear. I sat down. Slow down we think and then room got quiet. Zeno resumed posture of standing like vampire in shadow at ease. Primo gesture, primo mind. How bogus. Sat down and looked at the entire scene again. So then calculated that since 1997 an exact 87 years had passed. That was not too long, it seems. Then just calculated that since 1978 it had been 106 years. An impasse had been reached. Fuck."This is what you get for being so bookish. You never read Edgar Rice Boroughs, nor Ambrose Bierce nor the original Buck Rogers novel by Philip Francis Nowlan." From **your** Golden Age?" So sayeth Great Zeno.

"Nor fuckin' Jules Verne..." interrupted me.

"Shut up. So shadda fuckup!" yelled Zeno. She stood in shadows with **her crab eyes** closed so I looked away in anxiety. I decided to forget about starting shouting match. Ambience rules. Zeno stood in the same place motionless but in response opened her eyes. They were now a dull yellow which signalled "cool" so decided to give her replies an ear. Chill Out City. She explained that me was still her prisoner and we had some complex plans to discuss. None of it was negotiable. She then told me that because I had no hard data on any of the sciences nor really solid theology She had to take time to fill me in on background for mission she had trained me for. **They** were aware of my narrow education as a Journalism Major at Kent State plus what I had to do in the Ohio National Guard. Here was the gist of her message: Empress Zenobia and her problems in the Cosmos plus her solution:

Our basic situation began eons ago with that Great Experiment which is well known to you readers. That was back in the **Year 80 Million BC** or better yet, exactly 3086.4198 Sidereal Eons Before Christ. We call them Platonic Eons. Alice, you already know details of astronomy involved as well as some cartography. Aside from our charts on your Apple, which we still have, you also had some data pumped into your brain by us using neurology which you humans cannot achieve. The data itself is digital junk. Numeros and maps. But it takes up very small space even for your puny brain. Less than five Gigabytes, but good enough to convince some doctor at NASA in Houston that you have been given valid data by some advanced, make that very advanced, **ET persons unknown.** There is a certain bogus amount of secrecy involved, due to my cosmic war effort but that's what have to work with. It's my plan as Empress. We actually have this second team over in the USA who have been influencing both NASA and ESA to create a mission to the nearest star to Sol. That team is all **Neoforms,** like the one we left behind on Xiotan last week. Okay so far?

We will get to what we want from you in terms of action later. First, some more cosmic history: That Genome, Biome issue or whatever we may call it. That resulted in millions of years of warfare between two factions of us Eridan. Both of our tribes came from Xiodan, our Home World. Then we split into several and have in a long history, spread over the galaxy. But my faction, the Zenobian Tribe and the other one ruled by Empress Ariana, have been Ancient Enemies for that amount of time as you have been told. Our strategic point now is that three familiar stars are involved. Plus over 20 billion beings. Each Tribe has 10 billion as you can see. So we are evenly matched. The **really** vital fact now is that three planets are involved as well. That is hard to explain and happens to be a plan of mine. Their names are Terra, Xiotan and Nerus.

So Very important factoids are what me have. I need to win my wars so that my Citizens and me may survive! You must help us! Terra you know already. Except that some of your own data has have **point** to it. Those Men at NASA will give you great rewards for even the Earthish part of it. For now, Xiotan is our own name for our Home World. All Eridan came from it. So from now on we call it Xiotan. Okay? Nerus is the second planet of that target of your FIM. Which was and is in kindred spirit of Project Ozma. Well, First Interstellar Mission by NASA. It's in the Epsilon Indan System. Not, mind you, that you puny Space Cowboys wanna go there! No! Cream your jeans for Centauri System! Oh yeah! As if! Ha. Ha. Let me be Mother Mars. Even giant methane drinking crabs from Titan be under my command. No! Me shall divert your Future Man, as in Fred Biederman, Sentimental Journey to A.C. Okay? To save your Terran lives! And the Sacred Honor of NASA. So?

[**Editor:** Here the plot thickens. There be good plot here. Doth make for good reviews and nice royalties. Right?]

[**Tomas:** We hope so. Right?]

Was about to interrupt this far out monolog to ask what her plans for Race Mine could be but my emotions warned me. Why bother getting **God** angry? So, as in some New Age Concept, She is my Sister, my personal deity. Wow. Just go along with her and then we might become one too. Why not? Zoot allures. Allure of true godding. Hey! So me allowed Her to finish speech which was verbal. Me finally understand Her in emotional manner. She said, "You will travel there with twenty nine fellow astronauts and then see if any guerilla forces of our mutual enemy can be found there. Since Mankind comes in Peace, like your hero said in 1969, how can we go wrong?" Nixon, again? And me as American Nazi Bitch?

She added that any clash of that sort shall bring on the massive forces of Sirius Empire, for Zarcon. They shall help us win easy victory and my Tribe shall be given that entire system on a silver platter. **You get the planet.** It's the only one humans can live on. So was clear that three planets went through similar stages and Biomes. As we know, **Xiotan** was **the** only one of **our three planets** that was terraformed into its New Ecology. So me wish to keep it that way; Ariana wants to turn it back to its primal Biome, along with its Freon mix of air. That is her only way to maintain tribal loyalty. Such orbs have one major fact in common. Had perfectly even planes of orbit around their stars plus no tilt along their axis. Real stability of motion. **No seasons.** All had the same year of exactly 360 days. One minor detail: Because Terra is bigger, each day is 24 hours long. Nerus and Xiotan both had, and still have, a common day of 20 hours. With, natch, that Year of 360 days. Okay? So They say.

Both of our smaller two globes are still in the same precise cycles as the Great Architect has created them. Yet Terra has an anomaly which has happened long ago. Some meteor hit in Hudsons Bay or Gulf of Mexico which made your globe tilt by 23.5 degrees. Also your year is now 365 days. Which is an error we have to pay for and accounts for complexities of your native calendars. Then there's that shift in your Prime Meridian which is still a mystery for you and your **Race.** In other words, of our three planets, only Terra has seasons. Xiotan and Nerus do not. For that reason we now feel that your home planet is not needed for our purpose, which is that of very precise calibration for astrogation, etc. We are therefore not interested in it as a place to live on. Terra has too much water. Nor enough land to sustain Land animals. OK? **Your race can keep it.**

[**Editor:** We like this. A rather long Exposition. Of God knows what. We can accept this. Now carry on.]

For various reasons, the Prime Meridian or Point Zero for Terra was placed on a vertical line going thru Egypt long ago. We do not know who first charted your globe or when. Was it aliens or natives? Even we Eridan have no clue nor do we care. That Zero Latitude line cuts land masses in half. Your Equator then bisects it to quarter that. That small area seems to be at geographical center of Terran land masses. So about 3000 BC, King Cheops built Pyramids of Giza. We know of his motives cause we watching them build that great complex. Impressive for your Kind. But why? Only Real Point was if Me was Peter Sellers in some estupido 1962 comedy , is, am Me serious? Yes, as Me be Alien & dump upon Thee from Great Void. to serve as cartography marker. Like, as Beatnik say, get it? Your **scheisse** maps and calendars were to be calibrated according to such giant stone complex which functioned as **groovy** computer for your global culture. No doubt, Pharahos charged for the service and their astronomers became rich. So the ancient Line was not at Greenwich Village as it has been since 1688 AD, but right thru Top of Cheops and that was 30 degrees East of modern London. Also 33 from Stonehenge. "Okay so far?" She paused thank **God.** My head hurt.

Zenobia paused to let me offer a theory. I had one by chance. As reporter was aware of Islamic history and had for long deduced some facts behind those Turkish maps Von Daniken mentioned. Byzantines had for centuries kept the original cartography based on Giza as a military secret, as our First Evil Empire. When Islamic armies took Istanbul in **1453** may have wanted to keep it that way. However it was not to be, since refugees escaped to big cities of Italy where they used money and ancient know how to heat up the Rinascita, which was hot even then. Assume passively that Columbus got his hands on some of those charts. It was the major point he promptly followed thru with. That map of Antarctica may have been groovy, but in 1453 or even 1492, nobody wanted to go near that place.

Zeno laughed, "Too cold. Could stand on your South Pole naked and not feel it. Your ancestors cannot." I accepted her joke. So Piri Reis maps were not released to the media until after 1500. Then Zeno droned on. Where was I she said. [Now hear this: Turcoman Logos such say: Peri Reis was Deva. Some demon. Maybe an intellectual that was hired by the Emperor of Byzantium, then some Islamic Sultan to continue such science studies. Even then. **Peri** are same as us **Alien** Beings. From otros Star Systmemo, or maybe Io or Titan. Maybe some Archaea, living as Sulpholobus Extraphila Halocide. Like that pushy work of art. Zeno? Alice? N270? Noname? Was this a fucking joke? No! We are really so in Love with your Earth. OK? NASA like, as de real Exobilogy Labs, with Things on slabs? Ha! Try some captured as Prisoner Of War. From Titan: At pressure of 1.5 of Earth only. OK? Was OK. Beings that live there be giant crabs, as lounge about in vicinity of dark blue lakes of cyanide, with crystalline methane sand and silt, colored yellow, so cold, Kelvin 200 plus liquid, and warmer sand, warmer still, air of methane & ammonia gas mixture. Imagine: After 2004 NASA Huygens, Lander. My kind. Giant Archaea from moon of Saturn. My amigos. May be AAA+ soldat. Soldaten to conquer universe? You tell me. Oh yes. Politics and war. So anyway, there be three planets which have been vital to us. They used to be good for identifying points in Space. Some complex astrogation was based on them and so, they aided us in flying about the galaxy and also winning wars over the ages. Poetic, eh?

"Alice!" She said "Our enemies don't know about these facts and we plan to use it for strategic gain. For many reasons Eridan of my tribe have been ignoring life on Terra. Actually I still believe in tolerance and love anyway. The Galactic Council has given us a treaty that forces us to keep hands off Terra and is strictly enforced by the Sirians under Zarcon. Trader Flavius sells us hitech stuff so that keeps us in check as well. So recently we have a new policy. As Mankind expands into its own System, gets legal rights from Zarcon. We Eridan have to leave your system and our exodus is speeding up.

For instance in my present life was born on one Jovian moon. You call it Io. There are still masses of us living on Mars and the reason you cannot detect us yet is this: We can censor any telemetry coming from your probes and interfere with any telescopes. Its not hard. So as you can see, we picked you, Alice just because we needed some legal native of Terra and/or Sol System to become our ally. We have to attack all the billions of enemy Eridan still left in that new system which you have yet to visit, the Indan System and help us to conquer it ASAP. We need your help. So we have abducted you and trained you on our former Home World for skill and motivation. In your mind, body and soul you must become one of us! You gotta feel it. Okay?" I guess the above mass of data was to be pumped into my brain. Right?

Zeno paused and said, "Will allow questions, but just for now we have one more point. My name shall not be Zeno. Call me Galaxa instead. Now you may have the floor." What I did then was to get up and pace the floor in circles while casually talking. Was very calm not in spite of her monolog but because of it. All this cosmic debris amounted to an intense stream of galactic intrigue. Getting the message? Galaxa herself did not scare me as much as usual that day because was not concerned about her cosmic plots, which me was getting used to, but what she wanted of me on the personal level. What I really mean is that her presence tended to cause anxiety in others. It suited her role as Supreme Leader of some Empire. That was her role and we assumed that she had the physical & legal power to kill or torture her prisoners. But as time wore on from July of '78 to May 1979, got used to each other as long as we were physically in the same room; came by coincidence. Was casual afficiando kinda thing. Vicinity was the main thing with us two as our affinity grew. As them Spooks used to say, we needed face time in regular doses to like each other.

Well anyway, I had gotten used to her presence and so we grew ever closer as amigos but only, as I have said, when she was there. I could tell at once she was actually there in the same room. Phantoms soon become fake because you can tell. I also knew she had come here to comfort me in this part of my progress. I liked her ideas in terms of politics and strategy. It was becoming obvious that Her kind were natural allies of Mankind due to complicated intrigue out there in the Cosmos. Now that I for one finally understood her reasons, I finally agreed with her.

As the afternoon wore on. That phrase we learned as joke back in Ohio. Could tell it was so by observing Sol as it sank. We talked about business first. Then casually Galaxa said West was over there so that was Duisburg, a city across the Rhein. Another big city called Dortmund was fifty clicks to the East of our factory with vast urban sprawl between us. So we had to be in Essen. So what, I said. For that moment I had feelings of anxiety again and spent some time in the WC. Then came out and stood staring thru some window at my new outside world. As Brian Eno says, it was Another Green World. It was one of his LPs from his Ambient Music series. Brilliant. Wait. Feelings from similar albums were what I felt that day deep inside a valley in Western Europe. But this was 2084. Not 1978. Then I nervously surveyed the scene.

"People and cars out on the streets..." I said.

"Where?" asked Galaxa. Glare blinded me so turned away from the window. She told me it might be hard adjusting to this new world out there both in terms of time and space. Not only had we teleported but she had accelerated her fleet, which we needed as fighter escort, in some wide orbit that had made circles around Sol System at exotic velocity close to "cee" which allowed us to cover 106 years within days. Such figures had to be very rough for obvious reason. Had the same gut reaction of galloping panic which overcame me one year ago when I first awoke in Suite 6610. It had been a rush bordering on psychotic. This is why Galaxa had to baby me with some personal attention and all those luxuries in my yuppie tower. Also why we had landed in the similar biome of Rhine Valley. It was boiling hot outside and usually some real urban smog covered the sky, but today we only had a light haze. Inversion layer of air trapped in a bowl which we had here in Germany as well. There were some palmetto, citrus trees and figs among the plants outside.

It was not only the old greenhouse effect, as in just another century into the future, but also it had been close to tropic here anyway. Some amigos of mine had traveled thru Europe in 1970 and other times. Heard all about it. "What happens if we were to walk outside? Would we get arrested?" That sounded absurd. A question like that should sound stupid under any circumstances. But we were not only in my own society as de facto aliens but in the future of Mankind. Was actually used to aliens and by now afraid of my own kind! It got on my nerves. "Hey," I said "wait one minute. What was that about me not surviving past - what year was it? 1995? Ninety what? Seem to catch some snarky comment."

"I was suggesting that we may have saved your life by taking you out of those risky times you were living in. You know; social problems, money and lack of? Crime and even nuclear war? You know what I am getting at?"

"Your tone is smarmy and very unserious. Was that some song by Temptations? Rockets to the moon? Sound of soul? Hippies running to the hills? World War III? You look like Tina Turner in that skimpy gear and have the audacity to tell me of how you came down to save poor me of four billion poor beings? Piss off."

"Was about to tell you that it should be six billion by 1997 which is the year World War Three really broke out. Or maybe only five gigas. Who knows? Just made an estimate to prove that your world might have population growth to excess. To justify abduction. You know?" I said almost with a laugh "Yeah, I know alright. And that might of yours was too weak for words. None of us care. Okay?" Galaxa played the Innocent again so I launched into some rant about how not one citizen of Terra had any concern about the future. Not Personal nor collective. How many billion of us? It did not matter. She did not buy it. Even argued that, since my Kind were so selfish by nature it had been a favor. Really. So then I brought up Nixon and his so called enemies the Peaceniks. All that protest from 1965 to 1974. That era? Does it sound childish to you? The way we grew up in my own little world or was it just in the media? The college I went to. Keep in mind that, to be objective as a matter of my work, yet tended to deal in stereotypes for my raw material. It was done on paper, then on tape and now with IBMs and Apples. Still, so what?

"Your point is moral zero, Zeno." I said. She scared me not. I felt hunger an boredom.

"Who is Tina Turner? Okay, so I know who Nixon is. Brezhnev and Mao Tsetung and Chu Enlai, who for some reason seems to deserve great respect. That sort of thing me do okay with. I guess we Eridan if we wish for any more support really shall have to cut the cornball stuff and Get Real. Like no more 007 Espion moves and you can see the way we have to relate to you." said Zeno.

"That gives me clue as to what you are as a person and your Race. Was aware from Day One of how your mind came across. Klunky. That goes for rather stiff persona like me. An academic in shorts and halter top. In the times I lived in we were more Victorian than you think. Like your personal IQ level cannot be more than 160 no matter what. Its only one brain like mine. Goddess thing be only some metaphor. Even verbally be no big deal."

"I was about to compare Self to Cleopatra or Zenobia of 250 AD once more. Now I know that Race should be Mankind. Words can change and I understand. Do we sound politically incorrect?" Said it was not that. Just the klunky attitude. Limited range of gestures and idioms to express even the simplest of emotions and abstractions. We could - even the dumbest of us could tell an alien at once. Even people like me by 1975 had that notion in mind. Well, in the back of my mind anyhow. So my little amigo knew about that buzz word too. Media wars, nuke wars and now other wars. Or whatever. The bottom line was this: Galaxa had no choice but to come across as 1920 Hollywood actress.

Real corny. They call it a Vamp. Sophia Lauren is as hip as can be. So fine. So what if I have to make like was in some jerkoff Space Opera? I had that spooky feeling even on Day One. She was only trying to humor me. Trying to help me absorb the effects of both that Space War of hers plus - and this part bugged me more, the effects of Jumping thru Space. I had this notion that she was personally as flipped out as anyone by the effects of it. They tend to pile up. Also she was no tougher than those jerks in her platoon. Neo form or not. I followed a hunch.

"We have one thing in common." I said to humor her some more and was getting hungry plus annoyed. After this point intended to call for pizza like real soon. Anything. So I said it was the fact that both of us organisms in our personal and genome natures, went by the motto "Form follows Function" and even that ditty came from the Golden Age. This whole "do" had the makings of an okay party. For obvious reasons we could not just sit casually in some sidewalk café as Euros do. But I intended to get comfy as our adventure proceeded. So within 20 minutes we actually had some food. First found fone which still worked. Then used that to order pizza, some Chinese food and plenty of Coca Cola. In case our water was polluted. By rads or what not. Actually one of her soldats had attended to meeting the pizza guy at some factory security gate, where that alien did not stick out at all. No sore thumbs here. Maybe she used makeup for her face. To be sure, a Neoform does not attract attention here on Terra mainly because they breathe oxygen but also due to the fact that it we were squatting in some

dystopian "scene" like an empty factory in European rust belt, of all things. Sat down at a table and began to chow down. One of the grunts had just carried it in. Wore usual black cotton uniform they all did and even had a Belgian FN assault rifle
slung over her shoulder. Galaxa told me while ate and drank with gusto that we had fifty soldat like her living in the rooms below all armed. All looked exactly like the other billions and still did nothing but commune with ESP. This last one had not spoken a word. They nodded to each other to be polite. What robots. Galaxa sat down as if to socialize and pulled some alien food from her briefcase. She had also lugged along some packs of stuff which included both NATO and alien weapons. Why not? Her mission was by any standards well planned. Just to break the ice said, "You know, I like this Hollywood Bullshit of yours. We may be good allies."

"Good. I like smarmy talk like this. In fact, why don't me play the part of some USSR soldat? That would be credible. No? Then we could do our Seventies stylish Movie Wars, Nuke Wars and other stuff, and not one of them bastards on this pale blue dot of yours will notice. Carl Sagan. Cosmos. Just some book we had one of our Anthro Division staff to read. Look it up. Intuition is what we are at. Very cool. There are limits to progress, growth and mental agility. Seventies, eighties and nineties. Very long time to wait. Alice, why wait so long? It happened by chance. We have no idea of why. We had some random suspicion it may have been by The Year 2000."

This **thing** sat across from me at some office table eating her own food which was a paste made of sulfur hydroxide, sodium and pure carbon powder. She was more agile and due to the reactive nature of her biochemistry, as some NASA dude would say, was in command of fifty slaves who were more passive even if they all had copper blood in common. Her skin was dark blue ranging from purple to indigo in parts. Sweated out water that had mild ammonia compound in it. Galaxa breathed Freon gases, not pure of course, while her slaves had our native oxygen. She smiled and joked around while she ate, much like we all would. It was a wee party for two.

"I have a joke for you" I said, "This dude told me that in Europe there are these dudes who are, get this, very well spoken, polite and always dress up like dandies. They are also well respected. What are they?"

"I call it cheap humor, Alice. So do they wear disco suits? Blue suits? How about leather jackets? They are either intellectuals, terries or maybe both. Whoever you were talking to was bit funny in his head. Now give me one slice of pizza and one coke bottle. We can easily consume your food but the reverse does not apply. Do you know that? Biochem, amigo." Asked Zeno how her team had figured out what would happen in our Terran history in advance. How was it done if time travel by our feasible method were unreversable? "Like how did you know war would occur in 1997?"

Galaxa had stopped eating and stared at me. She then said one word: "Extrapolation." I asked her to elaborate so she said that they had first made made their plans to kidnap some Terran to serve as proxy for their ancient and smooth, precise strategy at a certain point in time. Used Haigisms which are just another Pentagon buzzword. By 1978 it was clear that the Western World was heading towards the Right in politics. So they used that fact and yes they had expert sociology to prove it. Just as an aside it was Asimov who must have developed real Sociology back in 1940 with his Foundation series. Hari Seldon was doing just that, like in Sociology, with his Mass Psychology method. So as you see, this being the Near Future we are in now, and me coming from the Very Far Future, we can almost emulate Hari Seldon. Then she stopped talking to eat and practice quiet wool gathering.

She had done it again. Nobody had to interrupt her. Had just lost Her train of thinking. She resumed after some more food by saying they used a method called retro engineering by observing Terran culture by the decade. For example see that thing on your PC? Started in 1980 as "bulletin" board, then LANS, then they called it the Web, Net or Internet. You can use it to order food like this plus other uses. Anyway, by 1997 technology in the comsumer part of global society stopped advancing. They had five years of nukes. So they went from five billion to under 800 million by 2002. Then in the Treaty of Brussels they ended it finally. Came and went in random manner. Now your planet has two billion in population. So we observed and decided to land in 2084. We have already placed some of our soldats here in this factory in stages. For some reason your nation is putting more effort into War and Space. In other words, we may as well be in the Year 1984 literally. Nothing but some narrow parts of your society has gone forward. "And that is buying us time." Galaxa said. I asked her what we were to do that night and so forth, noting that it was getting darker outside. I felt secure with my resident military politico taking care of me sitting right there enjoying nosh. Giving me personal attention. All of us had weapons as needed. We were also secretly here which was a sign of good planning, although I was no expert in these things. I was even allowed to have my own arsenal plus uniforms. Galaxa told me I was to meet a certain local person in some bar in what locals call **die Citi** in slang. It means **downtown** in English. How cool. Who was it?

"Her name is Cindi Nord. That is very cliched but what can I do? is one girl we can use on our mission for various reasons. She has her degree in metallurgy which got her into the European Space Agency and it also helps that she was in local NATO Reserves. She lives in a squat here in the same factory actually. For your info."

"Does she know what we are up to?" I asked.

"No. And I want to keep it that way. She is only a normal Terran who was born here and grew up like you. But about one century in the future. She is an Eighties person as far as we can define it. You need a sidekick. Cindi was born in 2056 which makes her 28. Your generation. Not much has changed. You two share the same primitive mindset we need. Just smart and strong enough to hand us the talent we need as astronaut but not to be conceited yuppies and have I just insulted you?"

"Yes but I can ignore it. So when do we do to this bar of yours? It must be Nine PM by now. Sol is about to set." Galaxa said we was to drive us there in her own car, which she explained was just some armored car they had stolen from NATO. They were all to pose as US Army and wear fake uniforms. Carefully figured it out and decided to stall for time, it being relative. So then Galaxa gave me another 24 hours and postponed our meeting for the next night. That was okay by her since it gave them all more time. Then I asked her a pressing question. "Why don't we just teleport there?"

"It would look stupid in case we are seen by the public." She said that it's also hard for her to perform often, with large amounts of material and over any sustained period of time. It takes intense effort and that is what you have to realize. I am an advanced being Eons old by means of being reborn serially. Can also possess the bodies of others, but have to be in a trance to do that. So I am not a real god, nor as talented as that Michael Valentine Smith guy. I know you've read the book."

[Later on, in the evening.]

There is a place in the oldest part of Essen called "La Cave Disco" which had existed once in the 1970s in various cities in Western Europe as one discotheque out of many. It was called "La Cave" simply as a brand name like Coca Cola or Pepsi or any other brand. In fact it was a franchise. It was always in basement of some house downtown and had no live bands. Yet they played rock music, not what we during the 1970s had come to call "disco" as a distinct style which had evolved from various types of Black Music around 1968. They had strange form of decor which was hip to most young people because it made no sense. Had grey wall to wall carpeting on the walls and ceiling, black rubber floor cover as it was found in modern airports plus a steel sheet dance floor. Its tables and chairs were made of white plastic and looked Jet Age. Had posters of Easy Rider, Neil Armstrong on the Moon on its walls. It was simply hip. Their bar was small and it only had one server but that was okay because they had the same brands of booze prices and so forth as any regular bar. That sort of place was known to most tourists from North America in My time. I had in fact heard about it from some amigos of mine.

There existed in Essen, large city in the Rhine Valley of Germany, duplicates of La Cave Disco in 2084. However in the future social conditions had changed. Islam had nothing to do with this, since I was to discover soon that all of the world's religions had declined in influence and power. Some had no more burocracy left. No longer **organized we** could say. Some had gone underground or turned into street gangs which, had happened during the 1990s to great extent. And had all that caused World War Three? Of that we have no clue.

Suffice it to say that in the European Union, which by 2084 had existed as one Nation for 82 years, religion was almost dead. Alcohol was still legal but many regarded it as evil due to radiation poisoning. Only the young and healthy still went to bars in Europe after 1997. So when Galaxa and I appeared there on the next day at some time that Evening it was totally empty. This gave me one very spooky feeling but then I recalled that my amigos had not arrived there until the Watergate Era. The Freak Era was actually already dead in spirit but its music was there. It was still pumping out the same chords and rythms now as before so it was one small part of some bizarre Historical Theme Park as far as I could surmise. Yes. Me. Alice. We walked up to the bar. Galaxa allowed me to order two beers. Just draft for one Euro each. By the way inflation goes up and down like yoyo - and so we decided to talk about it. For a while. We discussed Life in a casual manner before getting down to issue of meeting **Cindi Nord** and possible other persona grata. **Le Euro buck est official by 2002 anyway. Suffice to say, that made The Continent one gross Orwellian superstate.** Twas really so in **my** Sumerian Alternate **** Universe, by 2002. So like it or not; may say about Seventies y a los Eighties; as well never ended. **Was never no big thing.** One funny alternative like any other. You see, **voices** in that weird lounge called out to me. Am not dead. Am Alice Roanoke. **This is not like, We Are The Dead.** This is not stuff like Yobs With Celfones. Brit Labor This is what we wanted. As a person me wanted this. Then two girls came in. They looked like two chicks from the Seventies but only so because they adhered to some class system. They were not real down and out types, like with ragged denim and skinny because they had to do dope of some kind. Not even needle freaks who were common. No overt signs of poverty nor wealth. 1978.

Here the Look: They had on tight nylon blouses with long sleeves and those were pure white. Then mini skirts of flimsy cotton that flared out at the hips and were pleated. Those were salmon hued with white polka dots. Those skirts started out at the waist, which were narrow and ended abruptly four inches below the crotch. That was, low for any era they portrayed. Off hand 1970. By 1973 they had to be a bit lower, like almost halfway between knee and crotch. They had **really** wide hips like mine. Their legs were long and so were their arms. On left wrists they had femmy watches. Pale of skin and had curly blonde hair that was down to the ass. Their hairdo was just parted in the middle. They were both exactly five foot ten. So what. They had pretty faces devoid of makeup and any expression. Stoners.

They appeared to be Tweedy Dum y Tweedy Dee. Ha Ha. Well then fuck you up wazoo. Me und Zwei Germanic Princesses stood mesmerized in front of the bar and Niemand in any way noticed 'em. That was good. Because there were only five Beings in De Disco. Okay. Including de Barkeep, we had five. Well to make matters easier, this barkeep was only dude here as some Natural Man and he casually served customers almost as if they did not exist. I mean this man was cool. I like him. Was suave. Much Occult. Cosmic. Not of this Earth. That was me. **Alice Roanoke.** Peace & Love. Me. So that me only affected. With drink in hand, quiet. Wanker. Indeed. Had no bouncer nor waiters. Twas so, on Me Mothers Grave, etc. Cool. But shitty city. Europe, in 1970 nor in 2084 had no Work Creation Plan. That suited us fine.

Then Galaxa claimed to be a **Regular** here. Said it was sign of **class.** Too much.

Galaxa was dressed as Empress. Gold titanium bikini, which was bikini of course and was in my US Army combat uniform. That attracted no attention either. The two local hippie chicks walked into Le Cave side by side in unison. Maybe they were narcs? Looked at their feet you see pale bare legs with feet in sandals with socks. Another Euro trait. You see it was clear the Early Italic soldat wore sandals with thick socks to protect skin from injuries. That was the secret to cowboy boots, sandals & any other macho footwear. You simply needed heavy socks. Very sure they had something under their skirts. Yes, their titties were there. Not large but nice. Both girls were about 18 and they must have been twins. So now to the point of who were they? Popped the question to Galaxa. She said they were our Terra Nagual and served as contacts to Diablos who worked for a cause. Ask why. "Alice," Galaxa said "Let me now just intro Renate und Konstantine. Last name Anon and twins, as you see. They are local office workers. They are our Girl Friday Team and as Terran as you. We have arranged also this Evening for Cindi herself to show up. She will be your costar, like in some Hollywood movie. Our topic tonite will be NASA meets ESA. Get it?"

"Clear as beer piss." quoth me.

"That was not nice." said Renate, "How about clear as the azure sky?" She said so in perfect English. I knew this would happen. You see, German is not, as you may think, the same as English but a real other language. But by 2084 had almost died out. These two also spoke French & English because being trilingual was average for Citizens of The Union. Renate and amigo sat down with us at the bar then had half liters of Mosel Wein each. Galaxa and I were already into some beers plus Seagram's whiskey chasers. Yes, the Eridan love alcohol. Renate sat to the right of me while Konstantine sat to my left. Then Renate started our talk by staying that she was an expert in Astrophysics while still normal girl. She was still only a teenager who had grown up locally, in Duisburg much like Cindi. She worked as teller in some local bank. True to form, Galaxa interjected some comment about the typical Terran of 2084 being a Real Renaissance Man, then added some stuff about how H.G. Wells himself was a "self" educated man. Then Renate went on casually. She explained that I had spent one full year on Xiotan learning "siddhi" as we know them. On Earth. Such as are familiar to Buddhist plus other ancient Terran cultures. Also had been okay with such skills anyway due to my unique persona. Had been chosen from three million Cleveland candidates by Empress herself. All this teleporting was risky in many ways. All of us had to settle down and relax here for a month before we were free to fly over to the USA to begin our mission there. If I were to go now, the stress would tear me apart. Atoms of my body may become unstable such as particles in orbit. We are dealing with quantum effects which even now are only understood by using quantum computers Galaxa and some of Her Komrad aliens have. May have. Boom.

"So what can I do?" I asked. Renate said that I was to spend one month in their Base One as in that factory busy meditating, exercising, reading books, smoking weed, boozing and other casual activity. Once awhile the two girls would teach her about advanced astronomy and physics in a relaxed way. They would tour the libraries and cafes of this area, just for a break. Renate said "For you to go and try to adapt to modern NASA mindsets and also the shock of seeing your own North American society after 106 years would freak you out. You are to filter into NASA and enter their FIM, First Interstellar Mission as First Pilot, which they even call Astrogator. We already know how corny and Golden Agey all this stuff sounds, Ms. Roanoke."

I drank some more Mosel and agreed that Astrogator was corny indeed. It came right out of Tom Corbett. "You're only as good as material you have to work with," I said. Was becoming drunk and very interested but also found it hard to concentrate. There was nicotine all over and loud rock came from speakers. Some young people came in to dance and drink but it was all low key. Almost too casual for even my taste. Pub crawlers drifted in and out. Most of these humans had some level of radiation poisoning, which had been there for three generations. It was more so over past the Neo Iron Curtain, now called Neutral Zone. I felt sorry for them but was glad that they did not notice us. Most of these disco types, and the average person here in The Union, dressed 1970s style. Like baggy skirts and trousers, mostly pastel colors and natural fibers but some teens had on leather and other greasy gear. In other words, none of us aliens attracted any attention. Modern Europe was indeed the right ambience us total outsiders needed to survive.

Renate talked to me about the same stuff Zeno had always talked about. She repeated that joke I had heard already on Xiotan, about how "A person can attain perfect happiness just inside his own small room. What does he need the whole wide world for?" When Zeno had told me that, it was close to our departure and intended as cruel humor. I did not take it as any serious philosophy. Wonder if this was the best I could get from this bunch in terms of The Meaning of Life, or anything Cosmic. What they had was their own sense of humor based on ridiculing Religion. That was their basic common character trait. Had noticed that some humans other than just me able to live and function tribally with aliens and even had same sense of humor. At least in some emotional way we could function. You see, back in the Seventies most of us had assumed that aliens had nothing in common with us at all. Therefore it was not possible to communicate in any way with them. That was our consensus back in those days. Made sense because it was still a Rational point of view founded on Science and even Marxism. In focused way, I could handle all of this. Had some relaxed talk with this strange Renate person for over two hours in this disco. Our privacy was total.

The noise made it so. I nursed my beer religiously if you pardon the pun. Also Renate and me sat down at one table in an alcove. That was more comfortable. The Beatles song "Come together over me" was playing. It was about somebody called Old Flatfoot who was apparently some sort of "messiah" in his heart. The line "Come together over me" repeated itself to a pleasant rolling beat as a trance built up. Had this goofy notion that this hippie, who was also an undercover police eventually was killed, as if he were intended as human sacrifice. After he had finally ended "six feet under" they all gathered over his grave for his funeral. Hence that line "...over me..." as cynical refrain to all messiahs. It was that sorta spooky notion I was always partial to. It was never expressed, only kept inside. I was good at silent observation like that. Finally after midnite that other girl, Connie, came over & sat down with us. I asked her what she and Zeno had been talking about for so long. She casually said it was technical detail, like logistics.

They had to keep their small operation here in Essen going by hiring locals. Mostly they were female. It was hard to do even if their staff was only fifty. So for instance they had to get some local currency an have some Terran open bank accounts. They usually paid these agents in gold or other goods which came from the Eridan economy of course. Connie said, with the same bland stare both had that "Me, and that alien friend of yours, just talked about the usual stuff most people discuss. Small talk. We even allowed that barkeep to join in. But he's a personal pal anyway. He even knows Galaxa is an alien." Renate then agreed. She admitted that the two of them differed in character yet looked alike. She again explained that she herself had more in common with both me and Galaxa as cosmic types who always got into the Core Belief of what we were doing. But nobody was listening to you two and that was good. We can easily maintain secrecy because to the average Earth person, our war effort would seem absurd and meaningless even if they were to find out. Then Connie said that we had to remain superficial. She said that all she had to offer was normal talk such as her work, since she was actually a bank teller. Renate was more of a "dreamer" also by far the most "connected" person to Galaxa basically because she was unemployed as Euro. And so, had to be supported in part by my Eridan Army. That idea struck me as strange because I had never seen myself as dreamer. So had only one reason for this opinion: I never talked about business to others in general. So that I have applied myself to work yet came across as some hippie loser anyway. That may be simply an image problem.

I do not speak German. So that I noticed once in a while German words kept creeping into the conversation. Was that a Freudian slip? We think so. Renate had often used the words "abgestumpft" and "aufgeklart" and also "salop" but then, salop was en francais. Had also turned into just another modern slang word within deutsch. Like "filou" which means My Relative or better put, Mafia in the real French tongue. See how we Euros run? Like in Lady Madonna by The Beatle. That band had a truly Rubber Soul which had turned to about $100 million for each of them by 1970. Personally I had nothing against them. But simply not my cuppa tea.

All that and more Renate had explained to me in two hours. Euro Kultur at the time was as salop as always. Thus emotionally turned off. Attenuated? It was so. It had come to pass by 2002 when our Total War had finally ended. That was the experience of both Hip and New Wave as sub cults. I have said enough. Renate and Connie got up suddenly at once and bade me "adieu" somehow. In fact they gave me French Leave. That is when you never say "goodbye" which for some reason it was so typical for Euros. As both of them turned their backs to me, Renate still remained as a physical person but her twin had just suddenly vanished. Like some phantom ally. Walked over to Zeno still sitting at the bar after watching Renate leave by the door as a real human being of flesh and blood. Then yelled at her "That was not funny! You trying to pull a fast one on me?" Zeno broke down into utter mirth. She ran out of the disco and then stood on the street casual with feet wide apart, as if to invite me to kick Her in her cunt which still hurts. That was silly for some posture to take so I relaxed and asked what it was all about.

She explained to me briefly that Renate was the only girl of those two who really existed as a physical, living Terran. She was an "organic" person who was of course a real Cro Magnon, which was the actual name for her species; same as Zeno's species was Eridan. and that some sturdy and short Terrans were once called Neanderthal. That was actually there for real genetic situation. There were a few other evil and obscure ideas in genome science that I will not mention, and neither did Zeno. Pointed out that Renate had this habit of creating demons who actually performed tasks for her, such as work in a bank to cover provide a cover. That was her own head trip. That is called overdoing things.

It was dark outside and getting darker. Could see some local citizens out there on the streets of down town Essen, which they still call die Citi, and I began to wonder why this was so. Street lights were all out for some reason but I soon guessed why. Zeno had turned them off with her mind to make it dark more than it should be. Also her vibes must be creating an aura of fear, which I can do as well. Slowly out of the shadows, more of her soldat emerged. Armed with modern NATO weapons. Then armored cars showed up. They were actually unmarked but painted camo green. Zeno got into one of them with me and yelled some commands at her driver. All were Eridan. "Where is Renate?" I asked. "She went home by subway. You can call her tomorrow."

"Where are we off to now?"

Zeno told me as we roared thru the empty streets in tank motorcade of many armored cars that we were all headed back to Base One. That auto plant, in other words. Suddenly then thought of Cindi Nord, that girl from ESA. When were we to be introduced? Zeno told me in few words that Cindi was not to meet me until after I had arrived in Houston, Texas. That was to be during next month. I was to take Nasair flight from Brussels direct to NYC, then to Houston itself so that I could be hired by NASA. My first contact was Doctor Robert McGill. "Who is he?" I asked. The Chief Psychology Officer of NASA, came her answer. "Bigtime VIP, as you can imagine. All has been arranged for you including fake ID. You can even take a side trip to Ohio just to see your home town. First we need to have you taken care of at in Houston, with real astronaut training. You will live downtown in the Astro Center..."

"Astrodome?" I interjected. "No, silly! I mean the new Astronomy Center on 1776 Navigation Boulevard which is a major street near the core. you've been to Houston before, as we know" Zeno said.

"I know, but not after 1977. That was my last time there and it was only for hours on my way to Atlanta, Tampa and Miami on some work I had to do for the papers."

"Things have changed. Houston now has over eight million and larger in area, but still has Mission Control and other stuff we need to use for our own war effort. You will live in that giant NASA complex along with 29 others that you will get to meet. Training as Mission Specialists for that FIM launch. Gonna be bigtime. VIP all the way. There will be two from European Space Agency and the rest of this crew of thirty from USA. You'll like all of em, Alice. I can guarantee this. It will be what you've always dreamed of. A far out and cosmic experience. Think of it: over a century in the future and chance to meet some really stimulating, smart sensitive human beings. But one warning: they do not know about me, other aliens nor my waging of war. Understand?" Finally I agreed with Galaxa, sealing my personal contract on the spot. As our tanks raced thru the streets of this European city we alien warriors waited and speculated. So ends my story.

END OF CHAPTER EIGHT

Next: Chapter Nine

NASA Rover model 2080
Hybrid diesel & electro

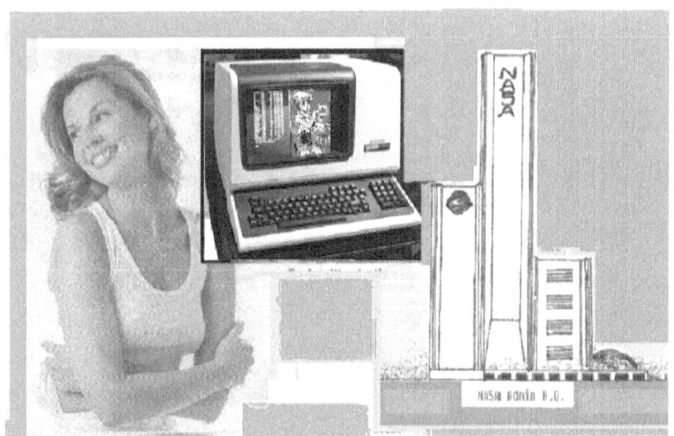

"Epilog"

This final page from our Epilog. After Cimora was placed under arrest by Empress Galaxa, her forces landed in many parts of Fontana Entheos. According to myth, where some sort of Fountain of Youth was now located. En choppers y truc, many Zenobian soldat came from **Safe** camps hidden out in **selva.** They did occupy this city and were welcome to local civilians as liberators. This part may sound corny. You can tell by its lack of **hipster jive.** Well, some of this was told me by Zeno herself. After our ruined dinner party we were told to go have some sleep. [Can I get a Pulitzer Price for this? S.V.P.]

Next day Zeno came into my room with this weird head fone device. It was attached to what looked like my own Panasonic tape deck, being an alien data storage machine utilizing what she called "Holocube" which was just one carbon crystal one inch to each side. This diamond cube fitted into a slot of her **Awareness** device. Hence, wired into Holy Akhashic Records database Itself as Flavius were later to explain. This worked as actual quantum computer. (Her term.) No human has ever seen such a thing. Directly into my brain it sent lots of data on astronomy. "Stuff that Mankind will never know," she said, "Which you will keep to yourself. Will help you navigate vast voids of Space between stars, velocity plus orbits within systems. That will allow you to teleport on your own, safely to some point on Earth. Later on, you might impress some experts at NASA in Houston with your **Advanced Fund** of memorized data on **this** part of our galaxy. Such may **fuck up your head a bit,** but that is part of our bargain. As it stands, you are **in** with my regime. **My Fucking Regime!** And we happen to be winners." She paused to let me soak it in. Effect.

Flavius came in to explain their motives. It was simple even as an epic tale. Long ago, some other star system had been property of Ariana's nation. That was the orange dwarf Epsilon Indus, part of Sirius Sector. But after this long period of warfare, our side had finally managed to destroy Really All Life upon one of its planets, the one we call Nerus. All life. Even down to microbes. Much like your own Mars. We can accuse Ariana of genocide. "This does not concern your Terra with its natives." Zeno added.

Flavius nodded. "In fact your nation is now our ally. We wanted you to join that first NASA mission to another star system which must happen about a century from now. That is my plan." Flavius said "She has my blessing as it were. It may be the proverbial good story." Sounds more like some goofy kinda fairy tale to me. Felt like saying. These cornballs had de nada for humor. It stank. Even felt like making some stupid comments about our last war over in, you know where? **That Thing with Nixon & Watergate.** My idea of Journalism - my work - is to be **objective.** Suddenly came to me that for always me to do is dig notion of being some kind of **galactic reporter.** It began to add up. Why not? They wanted me to fly off to some alien system and then just conquer it. These two freaks must think me have mind like ten year old. Even knew how such victory was achieved. By teleporting into the ships of some **Inferior Race.** Like home invasion. Then fly stolen ships to conquer all over. Why not? I was never to return to Xiotan again.

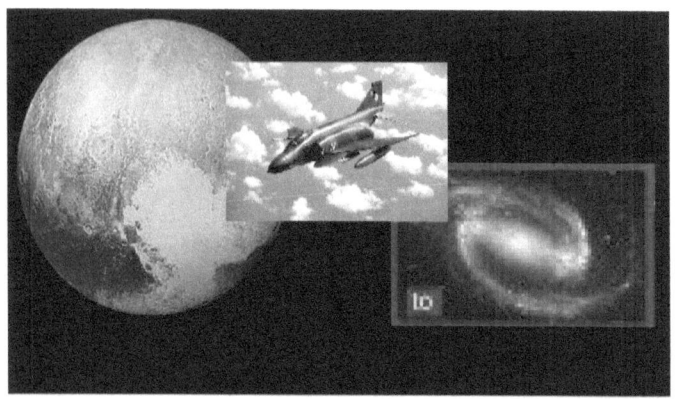

Conclusion:

In May of 1978 Alice was teleported back to Earth. She landed safely in the European city of Essen. Soon after, she was contacted in a "safe" house by her friend Galaxa. Alice is taken by Zeno Galaxa to secret meeting in Brussels. This takes place in the Head Office of the European Space Agency. There she is to meet and personally get to know the E.S.A. Mission Specialist Cindi Nord. Cindi has a degree in Metallurgy from University of Essen. Both will travel to NASA Headquarters in Houston, Texas. There they are to meet Dr. Robert McGill, Head of their Psychology Department. Both will be tested for their next major launch into space. They will join the crew of NASA's First Interstellar Mission called F.I.M. on a ship called Johnny Appleseed. In fact, Zeno herself made these arrangements. Some humans who work for ESA & NASA know that aliens live here & have infiltrated our society. In Book Two, Houston In 2084, we will find out why this complicated plan was set up.

END OF EPILOG

[**Editor's note:** Flavius & his Race the Omgal are total fiction. So are those Lizards from Planet Cetiwana, which appeared in an earlier book. Grey Guys & our various types of Eridan may exist. This series is a work of fiction.]

END OF BOOK ONE

Xiotan

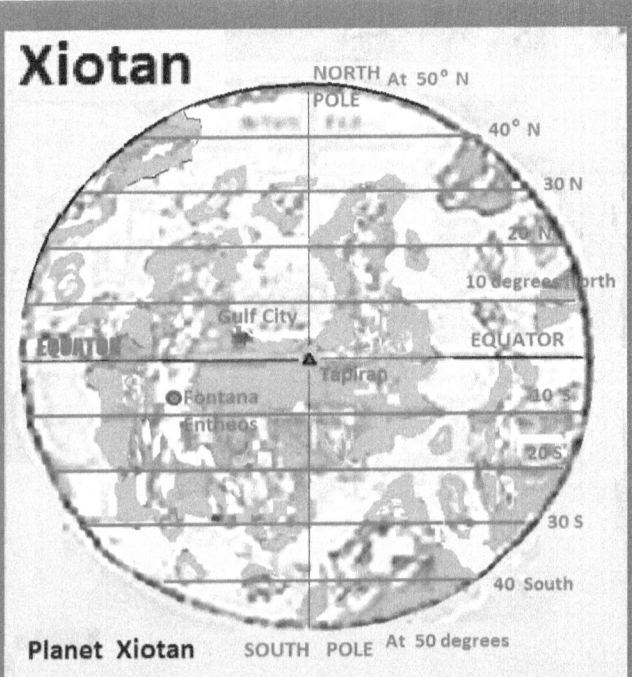

Planet Xiotan

Second in Epsilon Eridani System. At 6.5 AU from star. That is only 10.7 light years away. Planet has Earthlike air with oxygen. Diameter 7,500 miles. Grid divides into 100 degrees with exactly 75 miles per degree Latitude. Prime Meridian at Tapiran. Great Continent, seen here, is largest. Surface is 70% land and only 30% water. Oceans have salt water. Climate: Tropical humid.

Grid: Has 100 degrees from Pole to Pole. Each 10 d. is 750 miles, so that entire globe is 7,500 m. in diameter. Land surface 70% with oceans 30%. The Great Continent is on othere side.

Zafir the only moon.
Diameter: 1,500 miles.
Distance: 200,000 miles.

**Published Oct 2016 by
Tomas Londan and Amazon Inc.**

Isbn: 978-151-904-0909

This book is a work of fiction, and mention of any real person, group or place is coincidental. For mature readers only, containing profane words, nudity, drug abuse and scenes of extreme violence.

Below: Alternate cover.

Eridan Ozma
by Tomas Londan

Book One of the
Urbis Phobia Series

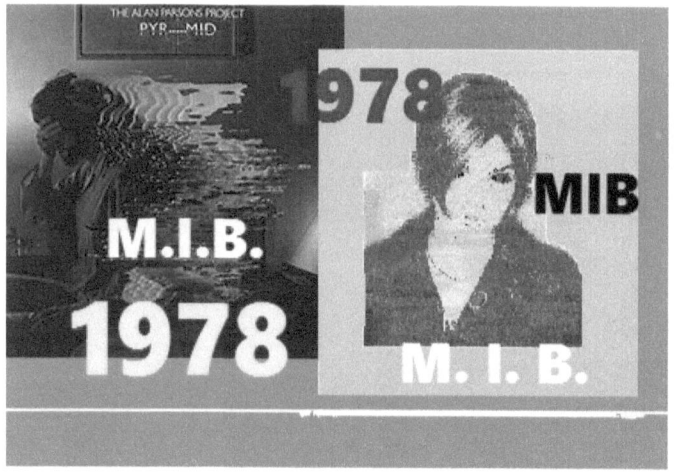

www.ingramcontent.com/pod-product-compliance
Lightning Source LLC
Chambersburg PA
CBHW031613210526
45464CB00004B/1550